The Practically Cheating Statistics Handbook

S. Deviant, MAT

StatisticsHowTo.com

info@statisticshowto.com

Printed in the United States of America

ISBN-13: 978-0578099095

Contents

Pre-Statistics

There are some elements from basic math and algebra classes that are must-knows in order to be successful in statistics. This section should serve as a gentle reminder of those topics.

1. How to Convert From a Percentage to a Decimal

"Percent" means how many out of one hundred. For example, 50% means 50 out of 100, 25% means 25 out of 100.

Step 1: *Move the decimal point two places to the left*.

Example: 75% becomes **.75%**.

Step 2: *Remove the "%" sign.*

.75% becomes **.75**.

2. *How to Convert From a Decimal to a Percentage*

Step 1: Move the decimal point two places to the right:

.75 becomes **75.00**

Step 2: Add a "%" sign.

75.00 becomes **75.00%**

Step 3 (Optional): You can drop the zeros if you want

75.00% becomes **75%**

3. How to Round to X Decimal Places

Sample problem: Round the number .1284 to 2 decimal places

Step 1: Identify where you want the number to stop by counting the number starting from the decimal point and moving right. One point to the right for the number .1284 would stop between the numbers 1 and 2. Two decimal places for the number .1284 will stop the number between 2 and 8:

.12|84.

Step 2: If the number to the right of that stop is 5 or above, round the number to the left up. In our example, .12|84, the 8 is larger than 5, so we round the left number up from 2 to 3:

.12|84 becomes **.13|84**.

Step 3: Remove any digits to the right of the stop:

.13|84 becomes **.13**.

Tip: Another way of looking at this is to take both numbers, either side of that stop mark I put in: 2|8 (we can forget the 4 at the end, as we're rounding way before that). Think of this as the number "28" and ask yourself: is this closer to 20, or closer to 30? It's closer to 30, so you can say .12|84, rounded to 2 decimal places, equals **.1300**. Then, you can just drop the zero.

4. *How to Use Order of Operations*

In order to succeed in statistics, you're going to have to be familiar with Order of Operations. The "operations": addition, multiplication, division, subtraction, exponents, and grouping, can be performed in multiple ways, and if you don't follow the "order", you're going to get the problem wrong. For example:

4 + 2 × 3

...can equal either 18 or 10, depending on whether you do the multiplication first or last:

4 + 2 × 3
6 × 3
18 (Wrong!)

4 + 2 × 3
4 + 6
10 (Right!)

Here's how to get it right every time with an acronym: PEMDAS.

The acronym PEMDAS actually stands for Parentheses, Exponents, Multiplication, Division, Addition, Subtraction, but you may easily remember it as "Please Excuse My Dear Aunt Sally". Let's say we have a question that asks you to solve the following equation:

1.96 + z

Where $z = \frac{\bar{X}-\mu}{s/\sqrt{n}}$

You are given:

$\bar{X} = 5$
$\mu = 3 \times 7$
$s = 5^2$
$n = 100$

Step 1: Insert the numbers into the equation:

$1.96 + z = 1.96 + ((5 - (3 \times 7)) / (5^2 / \sqrt{100}))$*
or
$1.96 + z = 1.96 + ((5 - (3 \times 7)) / 5^2 / 10))$**

Step 2: Parentheses. Work from the inside out if there are multiple parentheses. The inside parentheses is 3 × 7 = 21, so:

$1.96 + z = 1.96 + ((5 - 21) / (5^2 / 10))$

and there's only one operation left inside that first set of parentheses, so let's go ahead and do it:

$1.96 + z = 1.96 + ((-16) / (5^2 / 10))$

Next, we tackle the second set of parentheses.

Step 3: Exponents. $5^2 = 25$, so:

$1.96 + z = 1.96 + (-16 / (25 / 10))$

Step 4: There's no Multiplication in this equation, so let's do the next one: Division. Notice that there are two division signs. You may be wondering which one you should do first. Remember that we have to evaluate everything inside parentheses first, so we can't do the first division sign on the left until we evaluate (25 / 10), so let's do that:

$25 / 10 = 2.5$

$$\mathbf{1.96 + z = 1.96 + (-16/(2.5))}$$

Now we can do the second division:

$$\mathbf{-16 / 2.5 = -6.4}$$

$$\mathbf{1.96 + z = 1.96 + (-6.4)}$$

Step 5: Addition.

$$\mathbf{1.96 + (-6.4) = -4.44}$$

Step 6: Subtraction. In this case there is nothing more to do, so we're already done!

This should hopefully jog your memory–PEMDAS and Order of Operations is covered in basic math classes and you should be thoroughly familiar with the concept in order to succeed with statistics.

Notes:

* Notice that I placed parentheses around the top and bottom parts of the equation to separate them. If I hadn't, the equation would look like this:

$$\mathbf{z = 5 - 3 \times 7 / 52 / \sqrt{100}}$$

which could easily be interpreted (incorrectly) as this:

$$\mathbf{z = (5 - 3 \times 7 / 52) / \sqrt{100}}$$

so use parentheses liberally, as a reminder to yourself which parts of the equation go together.

** $\sqrt{100} = 10$. You don't need PEMDAS to evaluate simple square roots like this one.

5. *How to Find the Mean, Mode, and Median*

Step 1: *Put the numbers in order so that you can clearly see patterns.*
For example, let's say we have:

2, 19, 44, 44, 44, 51, 56, 78, 86, 99, 99

The **mode** is the number that appears the most often. In this case the mode is **44**—it appears three times.

Step 2: *Add the numbers up to get a total*. Example:

2 +19 + 44 + 44 +44 + 51 + 56 + 78 + 86 + 99 + 99 = 622

Set this number aside for a moment.

Step 3: *Count the amount of numbers in the series.*
In our example (2, 19, 44, 44, 44, 51, 56, 78, 86, 99, 99), we have 11 numbers.

Step 4: *Divide the number you found in Step 2 by the number you found in Step 3.* In our example:

622 / 11 = $56.\overline{54}$*

This is the **mean**, sometimes called the average.

If you had an odd number in Step 3, go to Step 5. If you had an even number, go to Step 6.

Step 5: *Find the number in the middle of the series.*

~~2,19,44,44,44~~, **51**,~~56,78,86,99,99~~
In this case it's **51**—this is the **median**.

Step 6: *Find the middle two numbers of the series.* Example:

~~1, 2, 5, 6~~, **7**, **8**, ~~12, 15, 16, 17~~

The **median** is the number that comes in the middle of those middle two numbers (7 and 8). To find the middle, add the numbers together and divide by 2:

(7 + 8) / 2 = **7.5**

Tip: You can have more than one mode. For example, the mode of

1, 1, 5, 5, 6, 6

is **1**, **5**, and **6**.

* An overline above decimal numbers means "repeating," so the number above is **56.54545454...** (going on forever).

6. *How to Find a Factorial*

Factorials are a concept usually introduced in elementary algebra. An exclamation mark indicates a factorial. Factorials are products of every whole number from 1 to the number before the exclamation mark. Examples:

2! = 1 × 2 = **2**

3! = 1 × 2 × 3 = **6**

4! = 4 × 3 × 2 × 1 = **24**

Most calculators have a factorial (!) button, but Google can also do the work for you!

Step 1: *Go to the search bar at Google.com*

Step 2: *Type in a factorial, such as 36!* (that's an exclamation point)

Step 3: *Press enter*

Google can also figure out more complicated factorials for you, such as

36! / (12 - 10)! 6!

However, make sure you put in parentheses and a multiplication sign (just as you would on any basic calculator):

36! / ((12-10)! × 6!) = $\mathbf{2.58328699 \times 10^{38}}$

7. How to Detect Fake Statistics

How do you know whether to trust results from a survey or not? Do you believe an egg company when it tells you 90% of consumers in a taste test preferred their eggs? How about if a voluntary survey of US marines showed overwhelming support for massive pay increases for military personnel? Sometimes it isn't enough to just accept the data as it is presented: dig a little deeper and you might uncover one of these common problems with statistics.

Step 1: Take a close look at **who paid** for the survey. If you read a statistic stating 90% of people lost 20 pounds in a month on a certain "miracle" diet, look at who paid for that survey. If it was the company who owns that "miracle" product, then it's likely you have what's called a **self-selection study**. In a self-selection study, someone stands to gain financially from the results of a trial or survey. You may have seen those soda ads where "90% of people prefer the taste of product X." But if the manufacturer of product X paid for that survey, you probably can't trust the results.

Step 2: Take a look at if the statistics came from a voluntary survey. A **voluntary response sample** is a sample where the participants can choose to be included in the sample or not. For example, if your professor was to send you an email with an invitation to comment on what you think of this book, then that would be a voluntary response sample. If it was a mandatory part of your course, then that would *not* be a voluntary response sample. Voluntary response samples are not suitable for statistics, because they carry a heavy bias toward people who have strong opinions (often negatives ones). In other words, students are more likely to respond to the above survey if they hate the textbook: the students who like the book will probably be less likely to respond.

Step 3: Look for the **faulty conclusion** that one variable causes another in the survey. For example, you might read a statistics that states unemployment causes an increase in corn production, because corn products (like high fructose corn syrup) are cheap, and therefore people are more likely to buy cheap foods when unemployed. But there may be many other factors causing an increase in production, including an increase in government subsidies for corn. Just because one factor is seemingly connected to another (correlation), that doesn't necessarily imply causation (that one caused the other)!

Step 4: Beware of **journal bias**. Journals are likely to report positive results (for example, a drug trial that had a positive outcome), rather than a drug trial that failed. Just because a journal publishes a positive result doesn't mean that there aren't other trials out there that reported a negative result.

Step 5: Make sure the **sample size** isn't too limited in scope. It's unlikely you can make generalizations about student achievement in the US by studying a single inner city school in Brooklyn. And it's unlikely you can make generalizations about American polling behavior by standing outside a polling booth in Ponte Vedra Beach, Florida. Just as inner city schools don't behave like every other school, an affluent neighborhood can't be used to generalize about the voting population. Also make sure the sample size is large enough: if your voting precinct contains 1 million voters, it's unlikely you'll get any good results from surveying 20 people.

Step 6: Watch out for **misleading percentages**. Unemployment may have "slowed by 50%," but if the unemployment rate was previously 100,000 new unemployment claims per month, that still means 50,000 people are joining the unemployed ranks every month.

Step 7: Beware of **precise numbers**. If a national survey reports that 3,150,023 households in the U.S. are dog owners, you might be inclined to believe that exact figure. However, it's highly unlikely (and almost impossible) that anyone would have seriously surveyed all of the households in the U.S. It's much more likely they surveyed a sample, and that 3,150,023 is an estimate (it should have been reported as 3 million to avoid being misleading).

There are many other examples of fake statistics. Newspapers sometimes print erroneous figures, drug companies print fake test results, governments present skewed statistics in their favor. The golden rule is: **question every statistic that you read**.

The Basics/Descriptive Statistics

1. How to Tell the Difference between Discrete and Continuous Variables

In an introductory stats class, one of the first things you'll learn is the difference between discrete and variable statistics. How to tell the difference between the two:

Step 1: *Figure out how long it would take you to sit down and count out the possible value of your variable.* For example, if your variable is "Temperature in Arizona," how long would it take you to write every possible temperature? It would take you literally forever:

50°, 50.1°, 50.11°, 50.111°, 50.1111°, ...

If you start counting now and never, ever, ever finish (i.e. the numbers go on and on until infinity), you have what's called a **continuous variable**.

If your variable is "Number of Planets around a star," then you can count all of the numbers out (there can't be infinity planets). That is a **discrete variable**.

Step 2: *Think about "hidden" numbers that you haven't considered.* For example: is time discrete or continuous? You might think it's continuous (after all, time goes on forever, right?) but if we're thinking about numbers on a wristwatch (or a stop watch), those numbers are limited by the numbers or number of decimal places that a manufacturer has decided to put into the watch. It's unlikely that you'll be given an ambiguous question like this in your elementary stats class but it's worth thinking about!

2. *How to Tell the Difference between Different Sampling Methods*

You'll come across many different types of samples in statistics: random samples, simple random samples, systematic samples, convenience samples, stratified samples, and cluster samples. Luckily, their names give away their meaning, so it's a straightforward process to determine what kind of sample you have.

Step 1: *Find out if the study sampled from individuals*. You'll find **random sampling** in a school lottery, where individual names are picked out of a hat randomly. A more "systematic" way of choosing people can be found in "systematic sampling," where every Nth individual is chosen from a population. For example, every 100th customer at a certain store might receive a "doorbuster" gift.

Step 2: *Find out if the study picked groups of participants*. For large numbers of people (like the number of potential draftees in the Vietnam War), it's simpler to pick people by groups (**simple random sampling**). Draftees were grouped by birth date.

Step 3: *Determine if your study contained data from more than one carefully defined group* ("strata" or "cluster"). Some examples of strata could be: Democrats and Republics, Renters and Homeowners, or Country Folk vs. City Dwellers. If there are two very distinct, clear groups, you have a **stratified sample** or a "cluster sample." If you have data about the individuals in the groups, that's a stratified sample. If you only have data about the groups themselves (you may only know the location of the individuals), then that's a **cluster sample**.

Step 4: *Find out if the sample was easy to get*. A classic example of **convenience sampling** is going to the mall and polling people who happen to walk by.

3. How to Tell the Difference between a Statistic and a Parameter

A statistic and a parameter are very similar. They are both descriptions of groups, like "50% of dog owners prefer Tasty Brand dog food." The difference is that statistics describe a **sample**, whereas a parameter describes an entire **population**.

For example, if you randomly poll voters in a particular election and determine that 55% of the population plans to vote for candidate A, then you have a **statistic**: you only asked a sample of the population who they are voting for, then you calculated what the population was likely to do based on the sample.

Alternatively, if you ask a class of third graders who likes vanilla ice cream, and 90% of them raise their hands, then you have a parameter: 90% of that class likes vanilla ice cream. You know this because you asked ***everyone* in the population**.

Here are some steps you take to be able to tell the difference between the two:

Step 1: *Ask yourself, is this obviously a fact about the whole population?* Sometimes that's easy to figure out. For example, with small populations, you usually have a parameter because the groups are small enough to measure:

- 10% of US senators voted for a particular measure. There are only 100 US Senators; you can count what every single one of them voted.
- 40% of 1,211 students at a particular elementary school got below a 3 on a standardized test. You know this because you have each and every student's test score.
- 33% of 120 workers at a particular bike factory were paid less than $20,000 per year. You have the payroll data for all of the workers.

Step 2: *Ask yourself, is this obviously a fact about a very large population?* If it is, you have a statistic.

- 60% of US residents agree with the latest health care proposal. It's not possible to actually ask hundreds of millions of people whether they agree. Researchers have to just take samples and calculate the rest, so this is a statistic.
- 45% of Jacksonville, Florida residents report that they have been to at least one Jaguars game. It's very doubtful that anyone polled in excess of a million people for this data. They took a sample, so they have a statistic.
- 30% of dog owners poop scoop after their dog. It's impossible to survey all dog owners—no one keeps an accurate track of exactly how many people own dogs. These data have to be from a sample, so it's a statistic.

When in doubt, think about the time and cost involved in surveying an entire population. If you can't imagine anyone wanting to spend the time or the money to survey a large number (or impossible number) in a certain group, then you almost certainly are looking at a statistic.

4. How to Tell the Difference between Quantitative and Qualitative Variables

In introductory statistics, it's easy to get confused when classifying a variable or object as quantitative or qualitative. Quantitative means it can be counted, like "number of people per square mile." Qualitative means it is a description, like "brown dog fur." How to classify those variables:

Step 1: *Think of a category for the items*, like "car models" or "types of potato" or "feather colors" or "numbers" or "number of widgets sold." The name of the category is not important.

Step 2: *Rank or order the items in your category*. Some examples of items that can be ordered are: number of computers sold in a month, students' GPAs or bank account balances. Anything with numbers or amounts can be ranked or ordered. If you find it impossible to rank or order your item, you have a qualitative item. Examples of qualitative items are "car models," "types of potato," "Shakespeare quotes."

Step 3: *Make sure you haven't added information*. For example, you could rank car models by popularity or expense, but popularity and expense are separate variables from "car model." If the item is "potatoes," it's qualitative. If the item is "number of potatoes sold," it's quantitative.

5. *How to Make a Frequency Chart and Determine Frequency*

If you are asked to determine a frequency in statistics, it doesn't just mean that you should just count out the number of times something occurs.

Step 1: *Make a chart for your data.* For this example, let's say you've been given a list of twenty blood types for incoming emergency surgery patients:

A, O, A, B, B, AB, B, B, O, A, O, O, O, AB, B, AB, AB, A, O, A

On the horizontal axis, write "frequency (#)" and "percent (%)". On the vertical axis, write your list of items. In this example, we have four distinct blood types: A, B, AB, and O.

	#	%
A		
B		
O		
AB		

Step 2: *Count the number of times each item appears in your data.* In this example, we have:

A appears 5 times

B appears 5 times

O appears 6 times

AB appears 4 times

Write those in the "number" column. This is your frequency.

	#	%
A	5	
B	5	
O	6	
AB	4	

Step 3: *Use the formula % = (f / n) × 100 to fill in the next column.* In this example, n = total amount of items in your data = **20**. A appears 5 times (frequency in this formula is just the number of times the item appears). So we have:

(5 / 20) × 100 = 25%

Fill in the rest of the frequency column, changing the 'f' for each blood type.

	#	%
A	5	25
B	5	25
O	6	30
AB	4	20

6. How to Find the Five Number Summary (Minimum, Q1, Median, Q3, and Maximum) By Reading a Boxplot

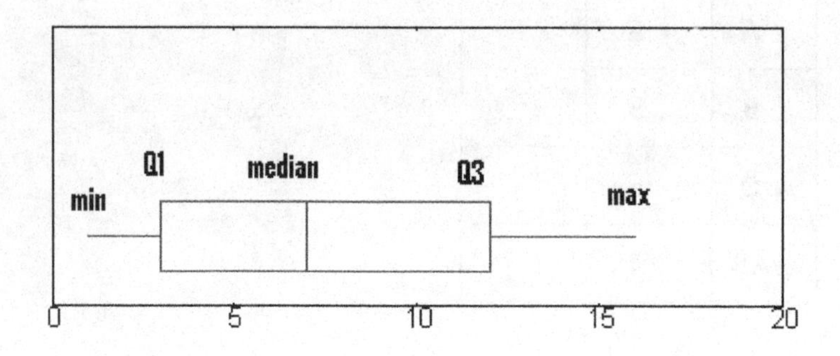

Step 1: *Find the Minimum.* The minimum is the far left hand side of the graph, at the tip of the left whisker. For this graph, the left whisker end is at approximately **.75**.

Step 2: *Find Q1.* Q1 is represented by the far left hand side of the box. In this case, about **2.5**.

Step 3: *Find the median*. The median is represented by the vertical bar. In this boxplot, it can be found at about **6.5**.

Step 4: *Find Q3*. Q3 is the far right hand edge of the box, at about **12** in this graph.

Step 5: *Find the maximum.* The maximum is the end of the "whiskers." In this graph, it's approximately **16**.

7. How to Find a Five Number Summary (Minimum, Q1, Median, Q3, and Maximum)

Step 1: *Put your numbers in order*. Example:

1, 2, 5, 6, 7, 9, 12, 15, 18, 19, 27

Step 2: *Find the minimum and maximum*. These are the first and last numbers in your ordered list. From the above example, the minimum is **1** and the maximum is **27**.

Step 3: *Find the median.* The median is the middle number of the list. In this case we have 11 total numbers, so the middle one is the sixth, which is **9**.

Step 4: *Place parentheses around the numbers above and below the median.* Not technically necessary, but it makes Q1 and Q3 easier to find.

(1, 2, 5, 6, 7), 9, (12, 15, 18, 19, 27)

Step 5: *Find Q1 and Q3.* Q1 can be thought of as a median in the lower half of the list, and Q3 can be thought of as a median for the upper half of data.

(1, 2, **5**, 6, 7), **9**, (12, 15, **18**, 19, 27)

In the case, Q1 is **5**, and Q3 is **18**.

8. How to Find an Interquartile Range

Sample problem: Find the Interquartile range of the numbers: 9, 5, 12, 19, 2, 18, 7, 15, 1, 27, 6.

Step 1: *Put the numbers in order:*

1, 2, 5, 6, 7, 9, 12, 15, 18, 19, 27

Step 2: *Find the median:*

1, 2, 5, 6, 7, **9**, 12, 15, 18, 19, 27

The median is **9**.

Step 3: *Place parentheses around the numbers above and below the median.* Not necessary statistically, but it makes Q1 and Q3 easier to spot.

(1, 2, 5, 6, 7), 9, (12, 15, 18, 19, 27)

Step 4: *Find Q1 and Q3.* Q1 can be thought of as a median in the lower half of the list, and Q3 can be thought of as a median for the upper half of list.

(1, 2, **5**, 6, 7), **9**, (12, 15, **18**, 19, 27)

Q1 is **5**, and Q3 is **18**.

Step 5: *Subtract Q1 from Q3 to find the interquartile range.*

18 – 5 = **13**

Note: This article tells you the steps, but if you want to find an IQR quickly, check out the Calculator section at the back of this book for a link to the online Interquartile Range Calculator.

9. How to Find an Interquartile Range on a Boxplot

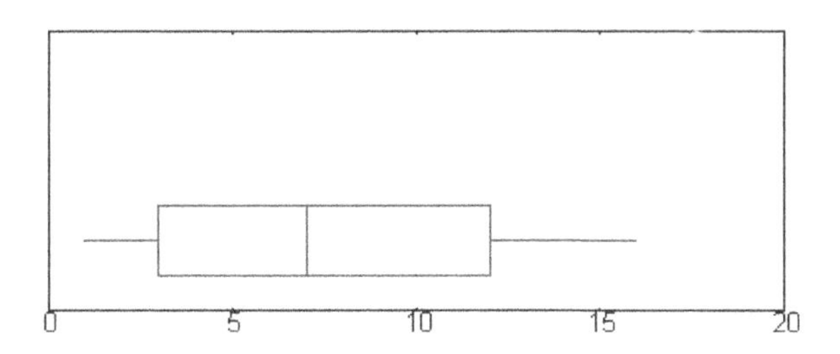

Step 1: *Find Q1.* Q1 is represented on a boxplot by the left hand edge of the "box." In the graph below, Q1 is approximately **2.5**.

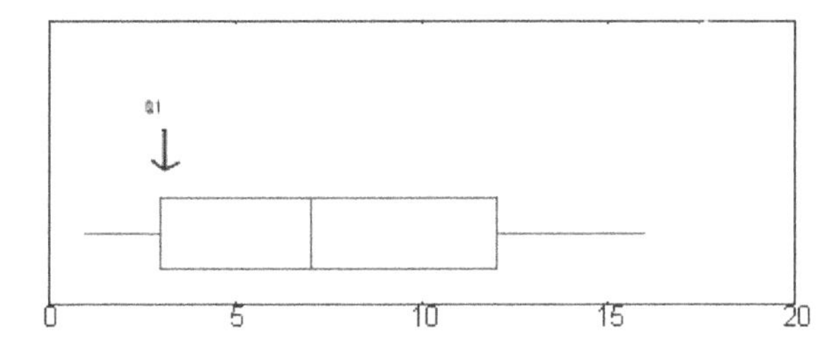

Step 2: *Find Q3.* Q3 is represented on a boxplot by the right hand edge of the "box." Q3 is approximately **12**.

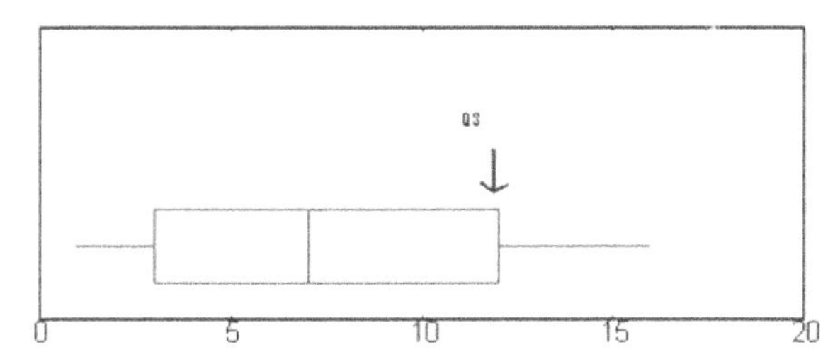

Step 3. *Subtract the number you found in Step 1 from the number you found in Step 3.* This will give you the interquartile range.

12 - 2.5 = **9.5**

10. How to Find the Sample Variance and Standard Deviation

The variance and standard deviations of samples are ways to measure the spread—or variability—of a sample or how far away from the "norm" a result is. In order to determine standard deviation, you have to find the variance first; standard deviation is just the square root of variance. The variance formula is:

$$\sigma^2 = \frac{\Sigma(X - \mu)^2}{N}$$

You could simply plug in the numbers. However, it can be tricky to use—especially if you are rusty on order of operations. By far the easiest way to find the variance and standard deviation is to use an online calculator (see the Resources chapter for links). However, if you don't have internet access when you are working these problems, here is how to find the variance and standard deviation of a sample in a few short Steps.

Step 1: *Add up the numbers in your given data set*. For example, let's say you were given a set of data for trees in California (heights in feet):

3, 21, 98, 203, 17, 9

Add them together:

3 + 21 + 98 + 203 + 17 + 9 = **351**

Step 2: *Square your answer:*

351 × 351 = **123,201**

...and divide by the number of items in your data set. We have 6 items in our example so:

123,201 / 6 = 20,533.5

Set this number aside for a moment.

Step 3: *Take your set of original numbers from Step 1, and square them individually this time:*

$$3 \times 3 + 21 \times 21 + 98 \times 98 + 203 \times 203 + 17 \times 17 + 9 \times 9$$

Add those numbers (the squares) together:

$$9 + 441 + 9604 + 41209 + 289 + 81 = \mathbf{51{,}633}$$

Step 4: *Subtract the amount in Step 2 from the amount in Step 3.*

$$51{,}633 - 20{,}533.5 = \mathbf{31{,}099.5}$$

Set this number aside for a moment.

Step 5: *Subtract 1 from the number of items in your data set.* For our example:

$$6 - 1 = \mathbf{5}$$

Step 6: *Divide the number in Step 4 by the number in Step 5:*

$$31{,}099.5 / 5 = \mathbf{6{,}219.9}$$

Step 7: *Take the square root of your answer from Step 6.* This gives you the standard deviation:

$$\sqrt{6{,}219.9} = \mathbf{78.86634}$$

11. How to Draw a Frequency Distribution Table

Sample Problem: The IQ scores of the children in a particular gifted classroom were recorded as 118, 123, 124, 125, 127, 128, 129, 130, 130, 133, 136, 138, 141, 142, 149, 150, 154. Draw a frequency distribution table of these data.

Step 1: Figure out how many classes (categories) you need. There are no hard rules about how many classes to pick, but there are a couple of general guidelines:

Pick between 5 and 20 classes. For the list of IQs above, we picked 5 classes.

Make sure you have a few items in each category. For example, if you have 20 items, choose 5 classes (4 items per category), not 20 classes (which would give you only 1 item per category).

Step 2: Subtract the minimum data value from the maximum data value. For example, our the IQ list above had a minimum value of 118 and a maximum value of 154, so:
154 – 118 = **36**

Step 3: Divide your answer in Step 2 by the number of classes you chose in Step 1.
36 / 5 = **7.2**

Step 4: Round the number from Step 3 up to a whole number to get the class width. Rounded up, 7.2 becomes **8**.

Step 5: Write down your lowest value for your first minimum data value:
The lowest value is **118**

Step 6: Add the class width from Step 4 to Step 5 to get the next lower class limit:
118 + 8 = **126**

Step 7: Repeat Step 6 for the other minimum data values (in other words, keep on adding your class width to your minimum data values) until you have created the number of classes you chose in Step 1. We chose 5 classes, so our 5 minimum data values are:
118
126 (118 + 8)
134 (126 + 8)
142 (134 + 8)
150 (142 + 8)

Step 8: Write down the upper class limits. These are the highest values that can be in the category, so in most cases you can subtract 1 from class width and add that to the minimum data value. For example:
118 + (8 – 1) = 125
118 – 125
126 – 133
134 – 142
143 – 149
150 – 157

Step 9: Add a second column for the number of items in each class, and label the columns with appropriate headings:

IQ	Number
118 – 125	
126 – 133	
134 – 142	
143 – 149	
150 – 157	

Step 10: Count the number of items in each class, and put the total in the second column. The list of IQ scores are: 118, 123, 124, 125, 127, 128, 129, 130, 130, 133, 136, 138, 141, 142, 149, 150, 154.

IQ	Number
118 – 125	4
126 – 133	6
134 – 142	4
143 – 149	1
150 – 157	2

Tip: If you are working with large numbers (like hundreds or thousands), round Step 4 up to a large whole number that's easy to make into classes, like 100, 1000, or 10,000.

12. How to Draw a Cumulative Frequency Distribution Table

In elementary statistics you might be given a histogram and asked to determine the cumulative frequency. Or, you might be given a frequency distribution table and asked to find the cumulative frequency. The method for both is the same.

Sample problem: Build a cumulative frequency table for the following classes.

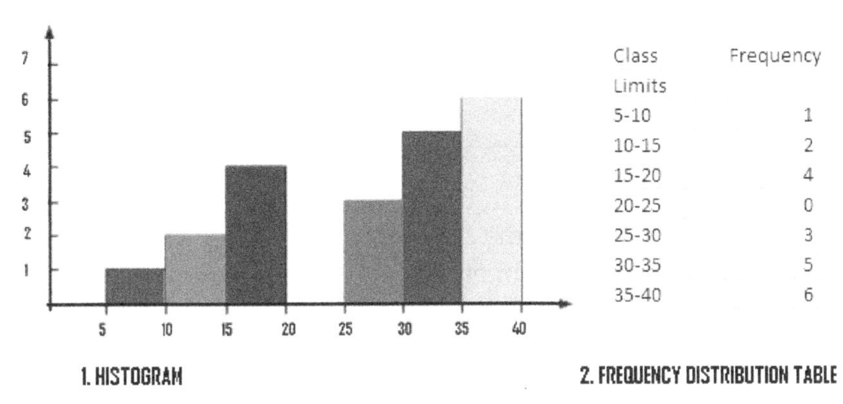

Class Limits	Frequency
5-10	1
10-15	2
15-20	4
20-25	0
25-30	3
30-35	5
35-40	6

1. HISTOGRAM

2. FREQUENCY DISTRIBUTION TABLE

Step 1: *If you have a histogram, go to Step 2. If you have a frequency distribution table (or both), go to Step 3.*

Step 2: *Build a frequency distribution table, like the one to the right of the histogram above*. Label column 1 with your class limits. In column 2, count the number of items in each class and fill the columns in as shown above. To fill in the columns, count how many items are in each class, using the histogram.

Step 3: *Label a new column in your frequency distribution table "Cumulative frequency" and compute the first two entries*. The first entry will be the same as the first entry in the frequency column. The second entry will be the sum of the first two entries in the frequency column.

Class	Frequency	Cumulative
5-10	1	1
10-15	2	3
15-20	4	
20-25	0	
25-30	3	
30-35	5	
35-40	6	

Step 4: *Fill in the rest of the cumulative frequency column.* The third entry will be the sum of the first three entries in the frequency column, the fourth will be the sum of the first four entries in the frequency column etc.

Class	Frequency	Cumulative
5-10	1	1
10-15	2	3
15-20	4	7
20-25	0	7
25-30	3	10
30-35	5	15
35-40	6	21

13. How to Calculate Chebyshev's Theorem

Sample problem: A left-skewed distribution has a mean of 4.99 and a standard deviation of 3.13. Use Chebyshev's Theorem to calculate the proportion of observation you would expect to find within two standard deviations (in other words, between -2 and +2 standard deviations) from the mean:

Step 1: *Square the number of standard deviations*:
$2^2 = \mathbf{4}$

Step 2: *Divide 1 by your answer to Step 1*:
1 / 4 = **0.25**

Step 3: *Subtract Step 2 from 1*:
1 – 0.25 = **0.75**

At least **75%** of the observations fall between -2 and +2 standard deviations from the mean.

Warning: As you may be able to tell, the mean of your distribution has no effect of Chebyshev's theorem! That fact can cause some wide variations in data, and some inaccurate results.

Probability

1. How to Construct a Probability Distribution

Sample problem: Construct a probability distribution for the following scenario: The probability of a sausage making machine producing 0, 1, 2, 3, 4, or 5 misshapen sausages per day are .09, .07, .1, .04, .12, and .02.

Step 1: *Write down the number of "widgets" given on one horizontal line.* In this case, we have "misshapen sausages."

Step 2: *Directly underneath the first line, write the result of the given function.* For example, the probability of the sausage machine producing 0 misshapen sausages a day is .09, so write ".09" directly under "0."

Number of misshapen sausages X	0	1	2	3	4	5
Probability P(X)	.09	.07	.1	.04	.12	.02

2. *How to Find the Probability of Selecting a Person from a Group or Committee*

Problems about picking people from groups or committees are as ubiquitous in probability and statistics as pajamas at a sleepover. It doesn't matter whether your problem has doctors and nurses, committees or boards, parents and teachers or some other group of people: the steps are all the same!

Sample problem: At a school board meeting there are 9 parents and 5 teachers. 2 teachers and 5 parents are female. If a person at the school board meeting is selected at random, find the probability that the person is a parent or a female.

Step 1: *Make a chart of the information given*. In the Sample problem, we're told that we have 5 female parents, 2 female teachers, 9 total parents and 5 total teachers.

Member	Female	Male	Total
Parent	5		9
Teacher	2		5

Step 2: *Fill in the blank column(s).* For example, we know that if we have 9 total parents and 5 are female, then 4 must be male.

Member	Female	Male	Total
Parent	5	4	9
Teacher	2	3	5

Step 3: *Add a second total to your chart* to add up the columns.

Member	Female	Male	Total
Parent	5	4	9
Teacher	2	3	5
Total	7	7	14

Step 4: *Add up the probabilities*. In our case we were asked to find out the probability of the person being a female or a parent. We can see from our chart that the probability of being a parent is 9/14 and the probability of being a female is 7/14:

$9/14 + 7/14 = \mathbf{16/14}$

Step 5: *Subtract the probability of finding both at the same time*. If we don't subtract this, then female parents will be counted twice, once as females and another as parents.

$16/14 - 5/14 = \mathbf{11/14}$

3. How to Find the Probability of an Event NOT Happening

Probability questions are often word problems that appear more difficult than they actually are. The trick is to identify the type of probability problem you have.

Step 1: *Choose one of the following Sample problems below that most closely matches your problem and go to the Step indicated.* Pay particular attention to the **bold** words when selecting.

"In a certain year, 12,942 construction workers were injured by falling debris. The top six cities for injuries were: New York (141), Houston (219), Miami (112), Denver (140), and San Francisco (110). If an injured construction worker is **selected at random**, what is the probability that the selected worker will have been injured in a city **other than** Houston, Miami, or Denver?" *If your problem looks like this, go to Step 2.*

"If the **probability** that a major hurricane will hit Florida this year is .49, what is the probability of a major hurricane **not** hitting Florida this year?" *If your problem looks like this, go to Step 5.*

"Twenty seven **percent** of American adults support Universal Health Care. If an American adult is chosen at random, what is the probability they **will not** support Universal Health Care?" *If your problem looks like this, go to Step 6.*

Step 2: *Add up the given cities.*

Houston (219) + Miami (112) + Denver (140) = **471**

Step 3: *Divide the number you found in Step 2 by the total number of people or items given.* In this case, we were told that 12,942 construction workers were injured:

471 / 12,942 = **.0364**

Step 4: *Subtract the number you found in Step 3 from the number 1:*

1 - .0364 = **.9936**

You're done!

Step 5: *Subtract the "not" probability from 1*:

1 - .49 = **.51**

You're done!

Step 6: *Convert the percentage to a decimal:*

Twenty seven percent = **.27**

Step 7: *Subtract the number in Step 6 from the number 1:*

1 - .27 = **.73**

You're done!

4. *How to Solve a Question about Probability Using a Frequency Distribution Table*

Sample problem: In a sample of 43 students, 15 had brown hair, 10 had black hair, 16 had blond hair, and 2 had red hair. Use a frequency distribution table to find the probability a person has neither red nor blond hair.

Step 1: *Make a table.* List the items in one column and the tally in a second column.

Type	Frequency
Brown	15
Black	10
Blond	16
Red	2

Step 2: *Add up the totals.* In the Sample problem we're asked for the odds a person will **not** have blond or red hair. In other words, we want to know the probability of a person having black or brown hair.

Brown = 15 of 43

Black = 10 of 43

15/43 + 10/43 = **25/43**

5. How to Find the Probability of A Simple Event Happening

Probability questions are word problems, and they can all be broken down into a few simple steps.

Sample problem: 50% of the families in the US had no children living at home. 22% had one child. 22% had two children. 4% had three children. 2% had four or more children. If a family is selected at random, what is the probability that the family will have three or more children?

Step 1: *Identify the individual probabilities and change them to decimals.* 4% of families have two children, and 2% have four or more children. Our individual probabilities are **.04** and **.02**.

Step 2: *Add the probabilities together.*

.04 + .02 = **.06**

Easy!

6. Finding the Probability of a Random Event Given a Percentage

Here's how to solve a problem that gives a percentage (i.e. 76% of Americans are in favor of Universal Health Care), then asks you to calculate the probability of picking a certain number (i.e. 3 people) and having them *all* fall into a particular group (in our case, they are in favor of health care).

Step 1: *Change the given percentage to a decimal.* In our example:

76% = **.76**

Step 2: *Multiply the decimal found in Step 1 by itself***.** Repeat for as many times as you are asked to choose an item, in or example, 3:

.76 × .76 × .76 = **.438976**

7. How to Find the Probability of Group Members Choosing the Same Thing

These probability questions give you a group, and ask you to calculate the probability of an event occurring for all of a certain number of random members within that group.

Sample problem: *There are 200 people at a book fair; 159 out them will buy at least one book. If you survey 5 random people coming out of the door, what is the probability they all will have purchased one book?*

Step 1: *Convert the data in the question to a fraction.* For example, the phrase "159 people out of 200" can be converted to **159/200**.

Step 2: *Multiply the fraction by itself. Repeat for however many random items (i.e. people) are chosen.* In this example, we have 5 people surveyed:

159/200 × 159/200 × 159/200 × 159/200 × 159/200 = **.3176**

8. How to Find the Probability of Two Events Occurring Together

Sample problem: 85% of employees have health insurance; out of those 85%, 45% had deductibles higher than $1000. How many people had deductibles higher than $1,000?

The key to these problems is the phrase "Out of this group" or "Of this group" which tells you that the events are dependent. You can solve these probability questions in only two Steps.

Step 1: *Convert your percentages of the two events to decimals.* In the above example:

85% = **.85**

45% = **.45**

Step 2: *Multiply the decimals from Step 1 together:*

.85 × .45 = **.3825**

9. How to Find the Probability of an Event, Given another Event

What these problems are asking you is 'given a certain situation, what is the probability that something else will occur?' These are called **dependent events.**

Sample problem: Using the table of survey results below, find the probability that the answer was no, given that the respondent was male.

Gender	Yes	No	Total
Male	15	25	40
Female	10	50	60
Total	25	75	100

Step 1: *Find the probability of both the events in the question happening together.* In our sample, question, we were asked for the probability of the answer no from a male. From the table, that is **25**.

Step 2: *Divide your answer in Step 1 by the total figure.* In our example, it's a survey, so we need the total number of respondents (100, from the table):

25 / 100 = **.25**

Step 3: *Identify which event happened first (i.e. find the independent variable).* In our example, we identified the male subgroup and then we deduced how many answered no, so "total number of males" is the independent event. The question usually gives this information away by telling you "**given that** the respondent was male..." (as in our question).

Step 4: *Find the probability of the event in Step 3.* In our example, we want the probability of being a male in the survey. There are 40 males in our survey, and 100 people total, so the probability of being a male in the survey is:

40 / 100 = **.4**

Step 5: *Divide the figure you found in Step 2 by the figure you found in Step 4:*

.25 / .4 = **.625**

10. How to Use a Probability Tree for Probability Questions

Drawing a probability tree (or a tree diagram) is a way for you to visualize all of the possible choices, and to avoid making mathematical errors.

Sample problem: An airplane manufacturer has three factories A, B, and C which produce 50%, 25%, and 25%, respectively, of a particular airplane. Seventy percent of the airplanes produced in factory A are passenger airplanes, 50% of those produced in factory B are passenger airplanes, and 90% of the airplanes produced in factory C are passenger airplanes. If an airplane produced by the manufacturer is selected at random, calculate the probability the airplane will be a passenger plane.

Step 1: *Draw and label lines to represent the first set of options in the question.* In our example, the options are the 3 factories, A, B, and C.

Step 2: *Convert the percentages to decimals*, and place those on the appropriate branch in the diagram. For our example:

50% = **.5**, 25% = **.25**

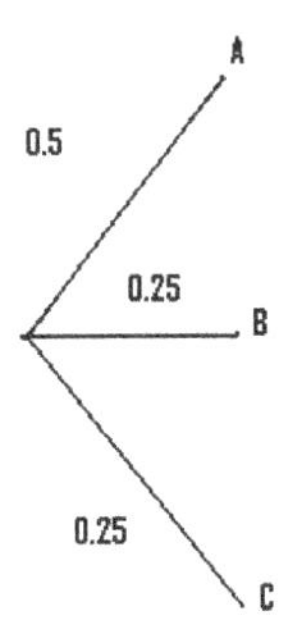

Step 3: *Draw the next set of branches*. In our case, we were told that 70% of factory A's output was passenger. Converting to decimals, we have .7 P ("P" is just my own shorthand here for "Passenger") and .3 NP ("NP" = "Not Passenger").

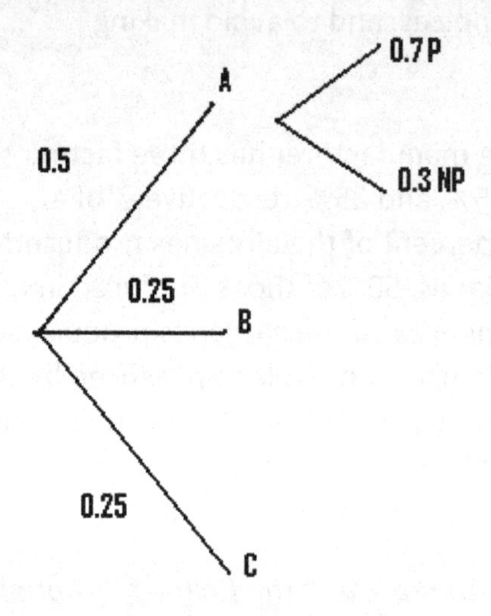

Step 4: *Repeat Step 3 for as many branches as you are given*.

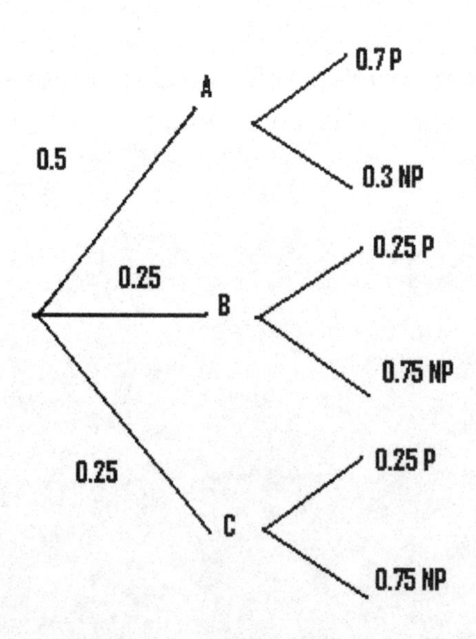

Step 5: *Multiply the probabilities of the first branch that produces the desired result together*. In our case, we want to know about the production of passenger planes, so we choose the first branch that leads to P.

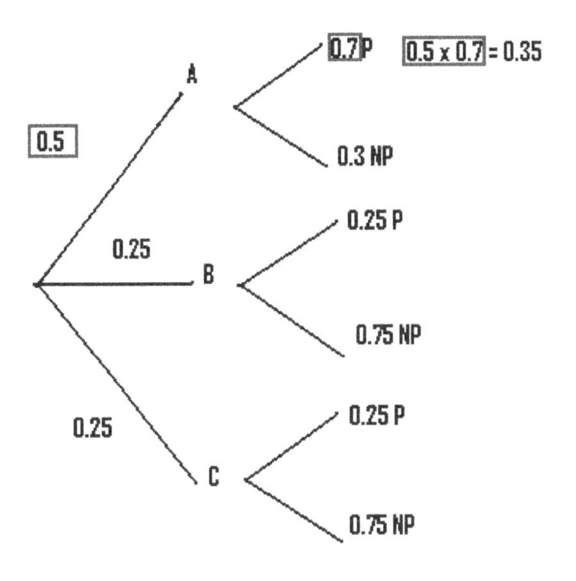

Step 6: *Multiply the remaining branches that produce the desired result.* In our example there are two more branches that lead to P.

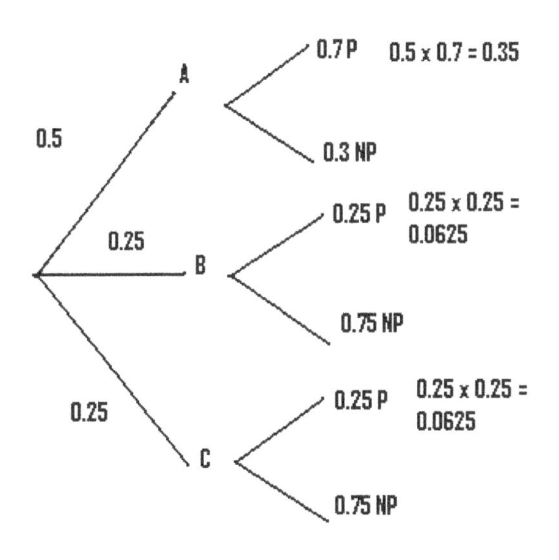

Step 7: *Add up all of the probabilities you calculated in Steps and 6:*

.35 + .0625 + .0625 = **.475**

11. How to Figure Out the Probability of Picking From A Deck Of Cards In Probability And Statistics

Questions about how to figure out the odds of getting certain cards drawn from a deck are ubiquitous in basic statistics courses. A common question is the probability of choosing one card, and getting a certain number card (i.e. a 7) or one from a certain suit (i.e. a club).

Step 1: *figure out the total number of cards you might pull.* Write down all the possible cards and mark the ones that you would pull out. In our example we've been asked the probability of a club *or* a 7 so we're going to mark all the clubs and all the sevens:

hearts: 2, 3, 4, 5, 6, **7**, 8, 9, 10, J, Q, K, A
clubs: **2, 3, 4, 5, 6, 7, 8, 9, 10, J, Q, K, A**
spades: 2, 3, 4, 5, 6, **7**, 8, 9, 10, J, Q, K, A
diamonds: 2, 3, 4, 5, 6, **7**, 8, 9, 10, J, Q, K, A

This totals 16 cards.*

Step 2: *Count the total number of cards in the deck(s)***.** We have one deck, so the total is **52**.

Step 3: *Express the result as a fraction***.** Divide Step 3 by Step 4:

16 / 52

* Be careful not to double count the 7 that is also a club. It would be easy to make the mistake of saying that there are four 7s in the deck, and 13 clubs, therefore the total is 17. But that's wrong because one of the 7s was already counted!

12. *How to Figure Out Mutually Exclusive Events*

Often in elementary statistics, you'll be asked to figure out if events are mutually exclusive.

Sample problem: If P(A) = .20, P(B) =.35, and (A U B) = .51, then are A and B mutually exclusive?

You would read that as: "If the Probability of A is 20%, the probability of B is 35%, and the probability of A given that B is true is 51%, then are A and B mutually exclusive?"

Step 1: *Add up the separate events (A and B)*:

.20 + .35 = **.55**

Step 2: *Compare your answer to the given "union" statement (A U B)*. If they are the same, the events are mutually exclusive. If they are different, they are not mutually exclusive.

In our example, .55 does not equal .51, so the events are **not mutually exclusive**.

13. *How to tell if an Event is Dependent or Independent*

Being able to tell the difference between a dependent and independent event is vitally important in solving probability questions.

Step 1: *Is it possible for the events to occur in any order?* If no (the steps must be performed in a certain order), go to Step 3a. If yes (the Steps can be performed in any order), go to Step 2. If you are unsure, go to Step 2.

Some examples of events that can clearly be performed in any order are:

- Tossing a coin, then rolling a die
- Purchasing a car, then purchasing a coat
- Drawing cards from a deck

Some events that must be performed in a certain order are:

- Parking and getting a parking ticket (you can't get a parking ticket without parking)
- Surveying a group of people, and finding out how many women are against gun rights (because you are splitting the survey into subgroups, and you can't split a survey into subgroups without first performing the survey).

Step 2: *Ask yourself, does one event in any way affect the outcome (or the odds) of the other event?* If yes, go to Step 3a, if no, go to Step 3b.

Some examples of events that affect the odds or probability of the next event include:

- Choosing a card, not replacing it, then choosing another, because the odds of choosing the first card are 1/52, but if you do not replace it, you are changing the odds to 1/51 for the next draw.

- Choosing anything and not replacing it, then choosing another (i.e. choosing bingo balls, raffle tickets)

Some examples of events that do not affect the odds or probability of the next event occurring are:

- Choosing a card and replacing it, then choosing another card (because the odds of choosing the first card are 1/52, and the odds of choosing the second card are 1/52)
- Choosing anything, as long as you put the items back

Step 3a: *You're done–the event is **dependent**.*

Step 3b: *You're done–the event is **independent**.*

14. Dice Rolling

Dice rolling questions are common in probability and statistics. You might be asked the probability of rolling a five and a seven, or rolling a twelve. By far the easiest (visual) way to solve these types of problems is by writing out a **sample space**.

A sample space is the set of all possible probabilities. For example, in order to know what the odds are of rolling a 4 or a 7 from a set of two dice, we would first need to find out what all the possible combinations are. We could roll a double one [1][1], or a one and a two [1][2]. In fact, there are 36 possible combinations.

Step 1: Write out your sample space. For two dice, the **36** different possibilities are:

[1][1], [1][2], [1][3], [1][4], [1][5], [1][6],
[2][1], [2][2], [2][3], [2][4], [2][5], [2][6],
[3][1], [3][2], [3][3], [3][4], [3][5], [3][6],
[4][1], [4][2], [4][3], [4][4], [4][5], [4][6],
[5][1], [5][2], [5][3], [5][4], [5][5], [5][6],
[6][1], [6][2], [6][3], [6][4], [6][5], [6][6]

Step 2: Look at your sample space and find how many add up to 4 or 7 (because we're looking for the probability of rolling one of those numbers).

[1][1], [1][2], **[1][3],** [1][4], [1][5], **[1][6],**
[2][1], **[2][2],** [2][3], [2][4],**[2][5]**, [2][6],
[3][1], [3][2], [3][3], **[3][4],** [3][5], [3][6],
[4][1], [4][2], **[4][3],** [4][4], [4][5], [4][6],
[5][1], **[5][2],** [5][3], [5][4], [5][5], [5][6],
[6][1], [6][2], [6][3], [6][4], [6][5], [6][6].

There are **9** possible combinations.

Step 3: Take the answer from Step 2, and divide it by the size of your total sample space from Step 1:

9 / 36 = .2s

You're done!

Binomial Probability Distributions

1. How to Determine if an Experiment is Binomial

Determining if a question concerns a binomial experiment involves asking four questions about the problem.

Sample problem: which of the following are binomial experiments?

- Telephone surveying a group of 200 people to ask if they voted for George Bush.
- Counting the average number of dogs seen at a veterinarian's office daily.
- You are at a fair, playing "pop the balloon" with 6 darts. There are 20 balloons, and inside each balloon there's a ticket that says "win" or "lose."

Step 1: *Are there a fixed number of trials*?

Question 1: yes, there are 200 (maybe binomial)

Question 2: no (not binomial)

Question 3: yes, there are 6 (maybe binomial)

Step 2: *Are there only 2 possible outcomes*?

Question 1: yes, they did or didn't vote for Mr. Bush

Question 3: yes, you either win or lose

Step 3: *Are the outcomes independent of each other*? In other words, does the outcome of one trial (or one toss, or one question) affect another trial?

Question 1: yes, one person voting for Mr. Bush doesn't affect the next person's response.

Question 3: I would say that the outcomes are probably independent, although you could argue that for each balloon popped, it increases or decreases your chance of an individual win since there are fewer unpopped balloons. Which brings us to...

Step 4: *Does the probability of success remain the same for each trial***?**

Question 1: yes, each person has the same chance of voting for MR. Bush as the last.

Question 3: yes, no matter how many balloons you pop, each has a 50% chance of being a winner.

2. *How to Find the Mean: Probability Distribution or Binomial Distribution*

Sample problem: A grocery store has determined that in crates of tomatoes, 95% carry no rotten tomatoes, 2% carry one rotten tomato, 2% carry two rotten tomatoes, and 1% carry three rotten tomatoes. Find the mean for the rotten tomatoes.

Step 1: *Convert all the percentages to decimal probabilities.*

95% = **.95**, 2% = **.02**, 1% = **.01**

Step 2: *Construct a probability distribution table.* (If you don't know how to do this, see the article on *constructing a probability distribution*).

Rotten Tomatoes X	0	1	2	3
Probability P(X)	.95	.02	.02	.01

Step 3: *Multiply the values in each column. In other words, multiply each value of X by each probability P(X).*
Referring to our probability distribution table:

0 × .95 = 0
1 × .02 = .02
2 × .02 = .04
3 × .01 = .03

Step 4: *Add the results from Step 3 together.*

0 + .02 + .04 + .03 = **.09**

3. *How to Find an Expected Value for a Single Item*

Sample problem: You buy one $10 raffle ticket for a new car valued at $15,000. Two thousand tickets are sold. What is the expected value of the raffle?

Step 1: *Construct a probability chart.* Put Gain(X) and Probability P(X) heading the rows and Win/Lose heading the columns.

	Win	**Lose**
Gain X		
Probability P(X)		

Step 2: *Figure out how much you could gain and lose.* In our example, if we won, we'd be up $15,000 less the $10 cost of the raffle ticket. If you lose, you'd be down $10. Fill in the data.

	Win	**Lose**
Gain X	$14,990	-$10
Probability P(X)		

Step 3: *In the bottom row, put your odds of winning or losing.* 2,000 tickets were sold, so you have a 1 in 2,000 chance of winning, and you also have a 1,999 in 2,000 chance of losing.

	Win	**Lose**
Gain X	$14,990	-$10
Probability P(X)	1/2,000	1,999/2,000

Step 4: *Multiply the gains (X) in the top won by the Probabilities (P) in the bottom row.*

$14,990 × 1/2,000 = **$7.495**
-$10 × 1,999/2,000 = **-$9.995**

Step 5: *Add the numbers from Step 4 together to get the total expected value.*

$7.495 + -$9.995 = **-$2.50**

In other words, when entering this raffle we can expect to lose $2.50, so it's not a wise idea to enter.

4. *How to Find an Expected Value for Multiple Items*

Sample problem: You buy one $10 raffle ticket, for the chance to win a car valued at $15,000, a CD player valued at $110, or a luggage set valued at $210. If 200 tickets are sold, what is the expected value of the raffle?

Step 1: *Construct a probability chart as in the "How to Find an Expected Value for a Single Item," except add columns for each possible prize*:

	Win	**Win**	**Win**	**Lose**
	Car	CD	Luggage	Cost
Gain X	$14,990	$100	$200	-$10
Probability P(X)	1/200	1/200	1/200	197/200

Step 2: *Multiply the gains (X) in the top won by the Probabilities (P) in the bottom row for each column:*

$14,990 × 1/200 = **$74.95**

$100 × 1/200 = **$.50**

$200 × 1/200 = **$1.00**

-$10 × 197/200 = **-$9.85**

Step 3: *Add the numbers from Step 2 together to get the total expected value.*

$74.95 + $.50 + $1.00 + -$9.85 = **$66.60**

In other words, when entering this raffle, we can expect to gain $66.60, so it makes sense to enter.

Note on the formula: The actual formula for expected gain in probability and statistics is

$$E(X) = \sum(X \times P(X))$$

What this says in English is "The expected value is the sum of all the gains multiplied by their individual probabilities."

5. *Standard Deviation: Binomial Distribution*

Most textbooks in elementary statistics will ask you to find the standard deviation in the binomial probability distribution chapter. It's time consuming because the variance and standard deviation formulas require a lot of arithmetic, but it's not difficult at all!

Sample problem: A sausage making machine occasionally produces misshapen sausages. The odds of the machine producing 0, 1, 2, 3, 4, or 5 sausages are .09, .07, .1, .04, .12 and .02 respectively. Find the variance and standard deviation.

Step 1: *Find the mean.*

Step 2: *Square the mean from Step 1.* For example, if your mean is .9, $.9^2$ = **.81**. Set this number aside.

Step 3: *Make a probability distribution chart.*

Number of misshapen sausages X	0	1	2	3	4	5
Probability P(X)	.09	.07	.1	.04	.12	.02

Step 4: *Square the top number, X, in each column and multiply it by the bottom number in the column, P(X).* For example $0^2 \times .09$ from the first column, $1^2 \times .07$. Repeat for all columns.

Step 5: *Add all of the numbers in Step 4 together.*

Step 6: *Subtract the number you found in Step 2 from the number you found in Step 5.*

Step 5 – Step 2 = ?

Step 7: *Take the square root of the number you found in Step 6.*

√Step 6 = ?

6. How to Work a Binomial Distribution Formula

The binomial formula can calculate the probability of success for binomial distributions. Often you'll be told to "plug in" the numbers to the formula and calculate. If you have a Ti-83 or Ti-89, the calculator can do much of the work for you. If not, here's how to break down the problem into simple steps so you get the answer right every time.

$$P(X) = \frac{n!}{(n-X)!\,X!} \times (p)^X \times (q)^{n-X}$$

Sample problem: 80% of people who purchase pet insurance are women. If 9 pet insurance owners are randomly selected, find the probability that exactly 6 are women.

Step 1: *Identify 'n' and 'X' from the problem*. Using our sample problem, n (the number of randomly selected items) is **9**, and X (the number you are asked to find the probability for) is **6**.

Step 2: *Work the first part of the formula:*

n! / (n - X)! X!

= 9! / ((9 - 6)! × 6!)

= **84**

Step 3: *Find p and q.* P is the probability of success and q is the probability of failure. We are given p: 80%, or **.8**. So the probability of failure is:

1 – .8 = **.2**

Or 20%.

Step 4: *Work the second part of the formula*:

p^X

$= .8^6$

= **.262144**

Step 5: *Work the third part of the formula*:

q^{n-X}

$= 2^{9-6}$

$= .2^3$

= **.008**

Step 7: *Multiply your answer from Step 3, 5, and 6 together:*

84 × .262144 × .008 = **.176**

In other words, there is a 17.6% chance that exactly 6 of the respondents will be female.

7. How to Use the Continuity Correction Factor

A continuity correction factor is used when you use a continuous function to approximate a discrete one; when you use a normal distribution table to approximate a binomial, you're going to have to use a continuity correction factor. It's as simple as adding or subtracting .5: use the following table to decide whether to add or subtract.

Given	In English, the probability that...	Correction
$P(X = n)$	X is exactly n	$P(n - .5 < X < n + .5)$
$P(X > n)$	X is greater than n	$P(X > n + .5)$
$P(X \geq n)$	X is greater than or equal to n	$P(X > n - .5)$
$P(X \leq n)$	X is less than or equal to n	$P(X < n + .5)$
$P(X < n)$	X is less than n	$P(X < n - .5)$

Step 1: *Identify the probability from your function.* Let's say you are given:

$P(X \geq 351)$

Step 2: *Replace the number with n:*

$P(X \geq 351)$

$P(X \geq n)$

Step 3: *Find the answer from Step 2* in the above table:

Given $P(X \geq n)$, use **$P(X > n - .5)$**

Step 4: *Put your number back into the right-hand equation*:

P (X > n – .5)

P (X > 351– .5)

Step 5: *Add or Subtract according to the equation:*

P (X > 351-.5)

P (X > 350.5)

8. How to Use the Normal Approximation to Solve a Binomial Problem

When ($n \times p$) and ($n \times q$) are greater than 5, you can use the normal approximation to solve a binomial distribution problem. This article shows you how to solve those types of problem using the continuity correction factor.

Sample problem: Sixty two percent of 12th graders attend school in a particular urban school district. If a sample of 500 12th grade children is selected from the district, find the probability that at least 290 are actually enrolled in that school.

Step 1: *Determine p, q, and n:*

p is defined in the question as 62%, or **.62**
q is 1 – p: 1 -.62 = **.38**
n is defined in the question as **500**

Step 2: *Determine if you can use the normal distribution*:

$n \times p$

$= 500 \times .62$

$= \mathbf{310}$

$n \times q$

$= 500 \times .38$

$= \mathbf{190}$

These are both larger than 5, so we can use the normal approximation.

Step 3: *Find the mean, μ by multiplying n and p:*

n × p

= **310**

Step 4: *Multiply the mean from Step 3 by q:*

310 × .38 = **117.8**

Step 5: *Take the square root of Step 4 to get the standard deviation, σ:*

√117.8 = **10.85**

Step 6: *Write the problem using correct notation*:

P (X ≥ 290)

Step 7: *Rewrite the problem using the* continuity correction factor. To learn how, see the section above, titled "How to Use the Continuity Correction Factor":

P (X ≥ 290)

P (X > 290 - .5)

= **P (X > 289.5)**

Step 8: *Draw a diagram with the mean in the center and your corrected n value from Step 7*. Shade the area that corresponds to the probability you are looking for. We're looking for X > 289.5:

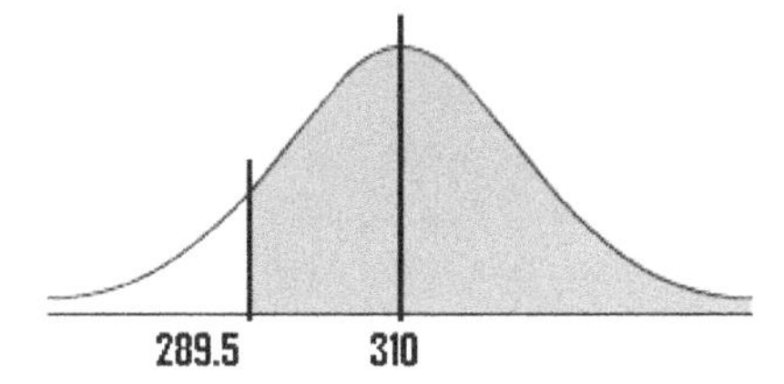

Step 9: *Find the z-value.* You can find this by subtracting the mean, μ, from the n value you found in Step 7, then dividing by the standard deviation, σ:

(289.5 – 310) / 10.85 = **-1.89**

Step 10: *Look up the z-value in the z-table (see Resources section at the back of this book):*

The area for -1.819 is **.4706**

Step 11: *Add .5 to your answer in Step 10 to find the total area pictured:*

.4706 + .5 = **.9706**

That's it! The probability is .9706, or **97.06%**.

Normal Distributions

1. How to Find the Area under the Normal Distribution Curve between 0 and a Given Z Value

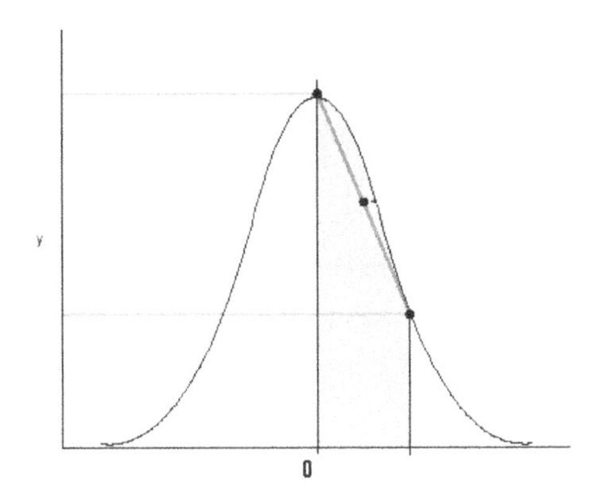

Sample problem: Find the area between 0 and .46.

Step 1: *Look in the z-table for the given z-score by finding the intersection.* See the Tables section in the back of the book for a link to an online z-table. In our example, we look up .46*.

0.4 is in the left hand column and .06 in the top row. The intersection is **.1772**, which is the answer.

z	0.00	0.01	0.02	0.03	0.04	0.05	0.06	0.07	0.08	0.09
0.0	0.0000	0.0040	0.0080	0.0120	0.0160	0.0199	0.0239	0.0279	0.0319	0.0359
0.1	0.0398	0.0438	0.0478	0.0517	0.0557	0.0596	0.0636	0.0675	0.0714	0.0753
0.2	0.0793	0.0832	0.0871	0.0910	0.0948	0.0987	0.1026	0.1064	0.1103	0.1141
0.3	0.1179	0.1217	0.1255	0.1293	0.1331	0.1368	0.1406	0.1443	0.1480	0.1517
0.4	0.1554	0.1591	0.1628	0.1664	0.1700	0.1736	0.1772	0.1808	0.1844	0.1879
0.5	0.1915	0.1950	0.1985	0.2019	0.2054	0.2088	0.2123	0.2157	0.2190	0.2224

* **Note**: Because the graphs are symmetrical, you can ignore the negative z-scores and just look up their positive counterparts. For example, if you are asked for the area of 0 to -.46, just look up .46.

2. *How to Find the Area under the Normal Distribution Curve in Any Tail*

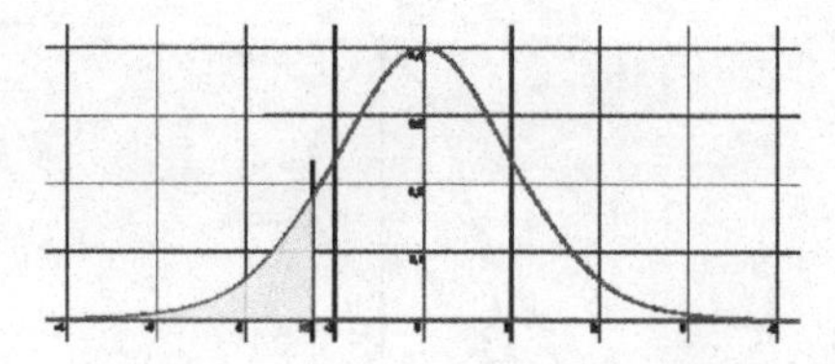

Sample problem: Find the area in the tail to the left of z = -.46.

Step 1: *Look in the z-table (see back of book for link) for the given z-value by finding the intersection.* In our example, we look up .46*.

.4 in the left hand column and .06 in the top row. The intersection is **.1772**.

z	0.00	0.01	0.02	0.03	0.04	0.05	0.06	0.07	0.08	0.09
0.0	0.0000	0.0040	0.0080	0.0120	0.0160	0.0199	0.0239	0.0279	0.0319	0.0359
0.1	0.0398	0.0438	0.0478	0.0517	0.0557	0.0596	0.0636	0.0675	0.0714	0.0753
0.2	0.0793	0.0832	0.0871	0.0910	0.0948	0.0987	0.1026	0.1064	0.1103	0.1141
0.3	0.1179	0.1217	0.1255	0.1293	0.1331	0.1368	0.1406	0.1443	0.1480	0.1517
0.4	0.1554	0.1591	0.1628	0.1664	0.1700	0.1736	0.1772	0.1808	0.1844	0.1879
0.5	0.1915	0.1950	0.1985	0.2019	0.2054	0.2088	0.2123	0.2157	0.2190	0.2224

Step 2: *Subtract the z-value you just found in Step 1 from .5:*

.5 - .1772 = **.3228**

* **Note**: Because the graphs are symmetrical, you can ignore the negative z-values and just look up their positive counterparts. For example, if you are asked for the area of a tail on the left to -.46, just look up .46.

3. How to Find the Area under the Normal Distribution Curve between Two Z-Scores on One Side of the Mean

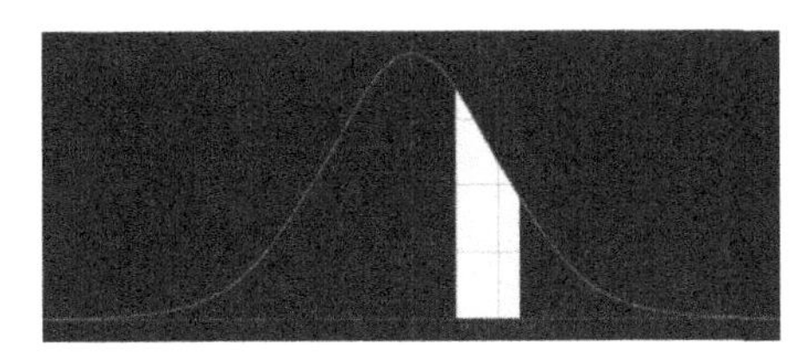

Sample problem: Find the area from z = -.46 to z = -.40.

Step 1: *Look in the z-table (see back of book for resources) for the given z-scores (you should have two) by finding the intersections.* In our example, we look up both .40* and .46*.

.4 in the left hand column and .06 in the top row; The intersection is **.1772**. .4 in the left hand column and .00 in the top row that intersection is **.1554**.

z	0.00	0.01	0.02	0.03	0.04	0.05	0.06	0.07	0.08	0.09
0.0	0.0000	0.0040	0.0080	0.0120	0.0160	0.0199	0.0239	0.0279	0.0319	0.0359
0.1	0.0398	0.0438	0.0478	0.0517	0.0557	0.0596	0.0636	0.0675	0.0714	0.0753
0.2	0.0793	0.0832	0.0871	0.0910	0.0948	0.0987	0.1026	0.1064	0.1103	0.1141
0.3	0.1179	0.1217	0.1255	0.1293	0.1331	0.1368	0.1406	0.1443	0.1480	0.1517
0.4	0.1554	0.1591	0.1628	0.1664	0.1700	0.1736	0.1772	0.1808	0.1844	0.1879
0.5	0.1915	0.1950	0.1985	0.2019	0.2054	0.2088	0.2123	0.2157	0.2190	0.2224

Step 2: *Subtract the smaller z-value you just found in Step 1 from the larger value.*

.1772 - .1554 = **.0218**

* **Note:** Because the graphs are symmetrical, you can ignore the negative z-values and just look up their positive counterparts. For example, if you are asked for the area of a tail on the left to -.46, just look up .46.

4. How to Find the Area under the Normal Distribution Curve between Two Z-Scores on Opposite Sides of the Mean

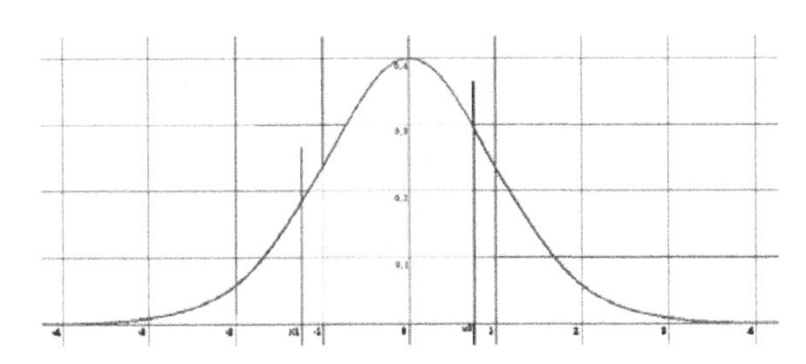

Sample problem: Find the area from z = -.46 to z = .16.

Step 1: *Look in the z-table* (see back of book for resources) *for the given z-scores (you should have two) by finding the intersections*. In our example, we, look up both .46* and .06.

.4 in the left hand column and .06 in the top row; the intersection is **.1772**. .0 in the left hand column, and .06 in the top row; the intersection is **.0239**.

z	0.00	0.01	0.02	0.03	0.04	0.05	0.06	0.07	0.08	0.09
0.0	0.0000	0.0040	0.0080	0.0120	0.0160	0.0199	0.0239	0.0279	0.0319	0.0359
0.1	0.0398	0.0438	0.0478	0.0517	0.0557	0.0596	0.0636	0.0675	0.0714	0.0753
0.2	0.0793	0.0832	0.0871	0.0910	0.0948	0.0987	0.1026	0.1064	0.1103	0.1141
0.3	0.1179	0.1217	0.1255	0.1293	0.1331	0.1368	0.1406	0.1443	0.1480	0.1517
0.4	0.1554	0.1591	0.1628	0.1664	0.1700	0.1736	0.1772	0.1808	0.1844	0.1879
0.5	0.1915	0.1950	0.1985	0.2019	0.2054	0.2088	0.2123	0.2157	0.2190	0.2224

Step 2: *Add both of the values you found in Step 1 together.*

.1772 + .0239 = **.2011**

* **Note:** Because the graphs are symmetrical, you can ignore the negative z-values and just look up their positive counterparts. For example, if you are asked for the area of a tail on the left to -.46, just look up .46.

5. How to Find the Area under the Normal Distribution Curve Left of a Z-Score

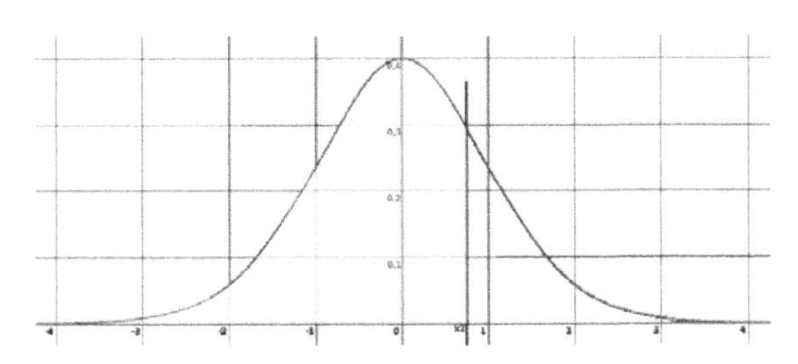

Sample problem: Find the area to the left of the z value .46.

Step 1: *Look in the z-table (see back of book for resources) for the given z-value by finding the intersection.* In our example, we look up .46*.

.4 in the left hand column and .06 in the top row. the intersection is **.1772**.

z	0.00	0.01	0.02	0.03	0.04	0.05	0.06	0.07	0.08	0.09
0.0	0.0000	0.0040	0.0080	0.0120	0.0160	0.0199	0.0239	0.0279	0.0319	0.0359
0.1	0.0398	0.0438	0.0478	0.0517	0.0557	0.0596	0.0636	0.0675	0.0714	0.0753
0.2	0.0793	0.0832	0.0871	0.0910	0.0948	0.0987	0.1026	0.1064	0.1103	0.1141
0.3	0.1179	0.1217	0.1255	0.1293	0.1331	0.1368	0.1406	0.1443	0.1480	0.1517
0.4	0.1554	0.1591	0.1628	0.1664	0.1700	0.1736	0.1772	0.1808	0.1844	0.1879
0.5	0.1915	0.1950	0.1985	0.2019	0.2054	0.2088	0.2123	0.2157	0.2190	0.2224

Step 2: *Add .5 to the z-value you just found in Step 1:*

.5 + .1772 = ***.6772***

* **Note:** Because the graphs are symmetrical, you can ignore the negative z-values and just look up their positive counterparts. For example, if you are asked for the area of a tail on the left to -.46, just look up .46.

6. How to Find the Area under the Normal Distribution Curve to the Right of a Z-Score

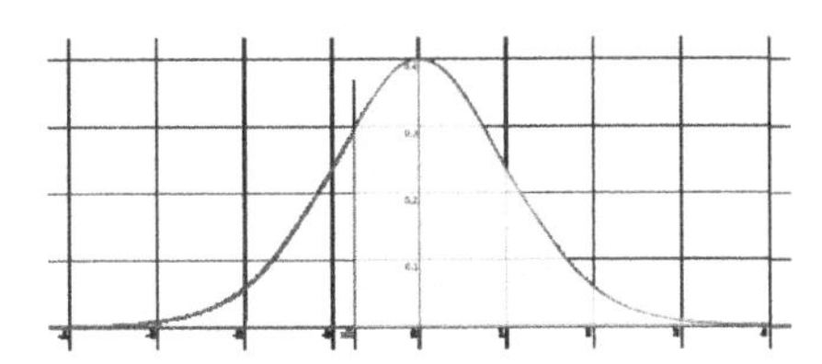

Sample problem: Find the area to the right of the z value -.46.

Step 1: *Look in the z-table for the given z-value by finding the intersection***.** In our example, we look up .46*.

.4 in the left hand column and .06 in the top row. The intersection is **.1772**.

z	0.00	0.01	0.02	0.03	0.04	0.05	0.06	0.07	0.08	0.09
0.0	0.0000	0.0040	0.0080	0.0120	0.0160	0.0199	0.0239	0.0279	0.0319	0.0359
0.1	0.0398	0.0438	0.0478	0.0517	0.0557	0.0596	0.0636	0.0675	0.0714	0.0753
0.2	0.0793	0.0832	0.0871	0.0910	0.0948	0.0987	0.1026	0.1064	0.1103	0.1141
0.3	0.1179	0.1217	0.1255	0.1293	0.1331	0.1368	0.1406	0.1443	0.1480	0.1517
0.4	0.1554	0.1591	0.1628	0.1664	0.1700	0.1736	0.1772	0.1808	0.1844	0.1879
0.5	0.1915	0.1950	0.1985	0.2019	0.2054	0.2088	0.2123	0.2157	0.2190	0.2224

Step 2: Add .5 to the *z-value you just found in Step 1*:

.5 + .1772 = **.6772**

* **Note**: Because the graphs are symmetrical, you can ignore the negative z-values and just look up their positive counterparts. For example, if you are asked for the area of a tail on the left to -.46, just look up .46.

7. How to Find the Area under the Normal Distribution Curve outside of a range (Two Tails)

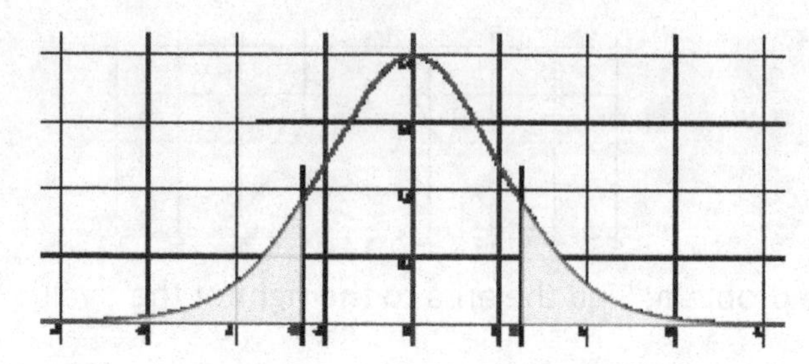

Sample problem: Find the area to the left of the z value -.46 and to the right of z value .46.

Step 1: *Look in the z-table for one of the given z-values by finding the intersection*. In our example, look up .46*.

.4 in the left hand column and .06 in the top row. The intersection is **.1772.**

z	0.00	0.01	0.02	0.03	0.04	0.05	0.06	0.07	0.08	0.09
0.0	0.0000	0.0040	0.0080	0.0120	0.0160	0.0199	0.0239	0.0279	0.0319	0.0359
0.1	0.0398	0.0438	0.0478	0.0517	0.0557	0.0596	0.0636	0.0675	0.0714	0.0753
0.2	0.0793	0.0832	0.0871	0.0910	0.0948	0.0987	0.1026	0.1064	0.1103	0.1141
0.3	0.1179	0.1217	0.1255	0.1293	0.1331	0.1368	0.1406	0.1443	0.1480	0.1517
0.4	0.1554	0.1591	0.1628	0.1664	0.1700	0.1736	0.1772	0.1808	0.1844	0.1879
0.5	0.1915	0.1950	0.1985	0.2019	0.2054	0.2088	0.2123	0.2157	0.2190	0.2224

Step 2: *Subtract the z-value you just found in Step 1 from .5:*

.5 - .1772 = **.3228**

Step 3: *Repeat Steps 1 and 2 for the other tail.* In our example, *the other tail also uses a z value of .46:*

.5 - .1772 = **.3228**

Step 4: *Add both z-values together.*

.3228 + .3228 = ***.6456***

* **Note**: Because the graphs are symmetrical, you can ignore the negative z-values and just look up their positive counterparts. For example, if you are asked for the area of a tail on the left to -.46, just look up .4

Central Limit Theorem

1. *Central Limit Theorem Problem Index*

A **Central Limit Theorem** word problem will most likely contain the phrase "assume the variable is normally distributed," or one like it. You will be given:

- A population (i.e. 29-year-old males, seniors between 72 and 76, all registered vehicles, all cat owners)
- An average (i.e. 125 pounds, 24 hours, 15 years, $196.42)
- A standard deviation (i.e. 14.4lbs, 3 hours, 120 months, $15.74)
- A sample size (i.e. 15 males, 10 seniors, 79 cars, 100 owners)

Choose one of the following questions and go to that section number:

Go to 2: I want to find the probability that the mean is **greater** than a certain number.

Go to 3: I want to find the probability that the mean is **less** than a certain number.

Go to 4: I want to find the probability that the mean is **between** a certain set of numbers either side of the mean.

2. *Central Limit Theorem: "Greater Than" Probability*

Sample problem: The there are 250 dogs at a dog show who weigh an average of 12 pounds, with a standard deviation of 8 pounds. If 4 dogs are chosen at random, what is the probability they have an average weight of more than 25 pounds?

Step 1: *Identify the parts of the problem*. Your question should state:

- mean (average or μ)
- standard deviation (σ)
- population size
- sample size (n)
- number associated with "greater than" ($\bar{X}$)

Step 2: *Draw a graph.* Label the center with the mean. Shade the area to the right of $\bar{X}$. This step is optional, but it may help you see what you are looking for.

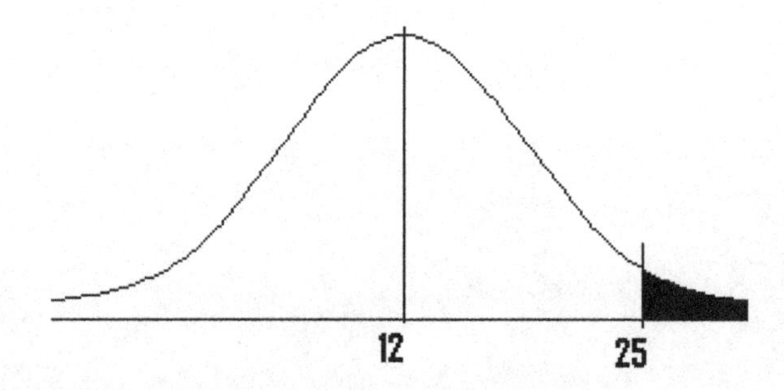

Step 3: *Use the following formula to find the z-value.* Plug in the numbers from Step 1.

$$Z = \frac{\bar{X}-\mu}{\sigma/\sqrt{n}}$$

If formulas confuse you, all this formula is asking you to do is:

Subtract the mean (μ in Step 1) from the greater than value ($\bar{X}$ in Step 1). Set this number aside for a moment.

Divide the standard deviation (σ in Step 1) by the square root of your sample (n in Step 1).

Divide your result from Step 1 by your result from Step 2.

$$Z = \frac{\bar{X} - \mu}{\sigma/\sqrt{n}}$$

$$Z = \frac{25 - 12}{8/\sqrt{4}}$$

$$Z = \frac{13}{8/\sqrt{4}}$$

$$Z = \frac{13}{8/2}$$

$$Z = \frac{13}{4}$$

Z = **3.25**

Step 4: *Look up the z-value you calculated in Step 3 in the* ***z-table*** *(see back of book for resources).*

Z value of 3.25 corresponds to ***.4994***

Step 5: *Subtract your z-score from .5:*

.5 - .4994 = ***.0006***

Step 6: *Convert the decimal in Step 5 to a percentage:*

.0006 = **.06%**

3. *Central Limit Theorem: "Less Than" Probability*

Sample problem: The there are 250 dogs at a dog show who weigh an average of 12 pounds, with a standard deviation of 8 pounds. If 4 dogs are chosen at random, what is the probability they have an average weight of less than 25 pounds?

Step 1: *Identify the parts of the problem*. Your question should state:

- mean (average or μ)
- standard deviation (σ)
- population size
- sample size (n)
- number associated with "less than" ($\bar{X}$)

Step 2: *Draw a graph.* Label the center with the mean. Shade the area to the left of $\bar{X}$. This step is optional, but it may help you see what you are looking for.

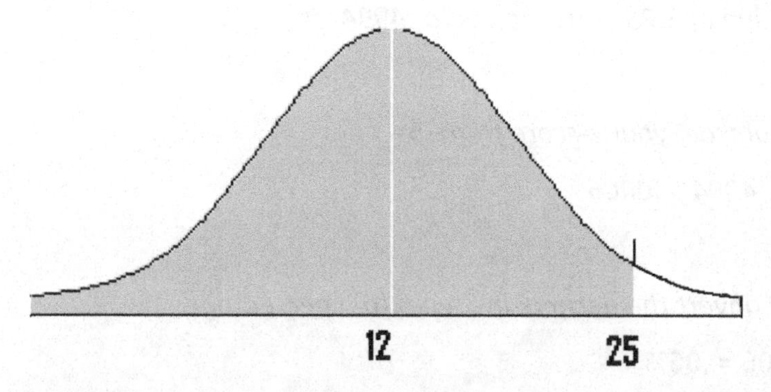

Step 3: *Use the following formula to find the z-value.* Plug in the numbers from Step 1.

$$Z = \frac{\bar{X}-\mu}{\sigma/\sqrt{n}}$$

If formulas confuse you, all this formula is asking you to do is:

Subtract the mean (μ in Step 1) from the greater than value ($\bar{X}$ in Step 1). Set this number aside for a moment.

Divide the standard deviation (σ in Step 1) by the square root of your sample (n in Step 1).

Divide your result from Step 1 by your result from Step 2.

$$Z = \frac{\bar{X}-\mu}{\sigma/\sqrt{n}}$$

$$Z = \frac{25-12}{8/\sqrt{4}}$$

$$Z = \frac{13}{8/\sqrt{4}}$$

$$Z = \frac{13}{8/2}$$

$$Z = \frac{13}{4}$$

Z = **3.25**

Step 4: *Look up the z-value you calculated in Step 3 in the z-table (see back of book for resources).*

Z value of 3.25 corresponds to ***.4994***

Step 5: *Add Step 4 to .5:*

.5 + .4994 = ***.9994***

Step 6: *Convert the decimal in Step 5 to a percentage:*

.9994 = **99.94%**

4. Central Limit Theorem: "Between" Probability

Sample problem: The there are 250 dogs at a dog show who weigh an average of 12 pounds, with a standard deviation of 8 pounds. If 4 dogs are chosen at random, what is the probability they have an average weight of greater than 8 pounds and less than 25 pounds?

Step 1: *Identify the parts of the problem.* Your question should state:

- mean (average or μ)
- standard deviation (σ)
- population size
- sample size (n)
- number associated with "less than" $\bar{X}_1$
- number associated with "greater than" $\bar{X}_2$

Step 2: *Draw a graph.* Label the center with the mean. Shade the area between $\bar{X}_1$ and $\bar{X}_2$.This step is optional, but it may help you see what you are looking for.

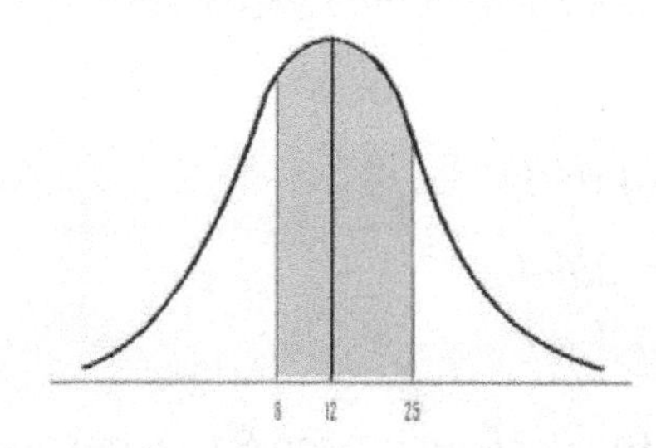

Step 3: *Use the following formula to find the z-values.* Plug in the $\bar{X}_1$ from Step 1:

$$Z = \frac{\bar{X} - \mu}{\sigma/\sqrt{n}}$$
$$Z = \frac{25-12}{8/\sqrt{4}}$$
$$Z = \frac{13}{8/\sqrt{4}}$$
$$Z = \frac{13}{8/2}$$
$$Z = \frac{13}{4}$$
Z = **3.25**

All this formula is asking you to do is:

a) Subtract the mean (μ in Step 1) from the greater than value ($\bar{X}$ in Step 1).
b) Divide the standard deviation (σ in Step 1) by the square root of your sample (n in Step 1).
c) Divide your result from *a* by your result from *b*.

Step 4: *Use the following formula to find the z-values.* Plug in the $\bar{X}_2$ from Step 1:

$$Z = \frac{\bar{X} - \mu}{\sigma/\sqrt{n}}$$
$$Z = \frac{8-12}{8/\sqrt{4}}$$
$$Z = \frac{-4}{8/\sqrt{4}}$$
$$Z = \frac{-4}{8/2}$$
$$Z = \frac{-4}{4}$$
Z = **-1**

Step 5: *Look up the z-value you calculated in Step 3 in the z-table*

*Z value of 3.25 corresponds to **.4994***

Step 6: *Look up the z-value you calculated in Step 4 in the z-table (see back of book for resources).*

Z value of 1 corresponds to ***.3413***

Step 7: *Add Step 5 and 6 together:*

.4994 + .3413 = **.8407**

Step 8: *Convert the decimal in Step 7 to a percentage:*

.8407 = 84**.07%**

Confidence Intervals

1. How to Find a Sample Size Given a Confidence Interval and Width (Unknown Standard Deviation)

Sample problem: 41% of Jacksonville residents said that they had been in a hurricane. How many adults should be surveyed to estimate the true proportion of adults who have been in a hurricane, with a 95% confidence interval 6% wide?

Step 1: *Using the data given in the question, figure out the following variables:*

- $z_{a/2}$: Divide the confidence interval by two, and look that area up in the z-table (see resources at back of book):

  .95 / 2 = .475

  The closest z-score for .475 is **1.96**

- E (margin of error): Divide the given width by 2.

  6% / 2

  = .06 / 2

  = **.03**

- $\hat{p}$: use the given percentage.
  41% = **.41**
- $\hat{q}$: subtract $\hat{p}$ from 1.
  1 - .41 = **.59**

Step 2: *Multiply* $\hat{p}$ *by* $\hat{q}$. Set this number aside for a moment.

.41 × .59

= **.2419**

Step 3: *Divide* $z_{\alpha/2}$ *by E.*

1.96 / .03

= **65.3333333**

Step 4: *Square Step 3*:

65.3333333 × 65.3333333

= **4,268.44444**

Step 5: *Multiply Step 2 by Step 4:*

.2419 × 4268.44444

= 1,032.53671

= **1033** people to survey

2. How to Find a Sample Size Given a Confidence Interval and Width (Known Standard Deviation)

Sample problem: Suppose we want to know the average age of the students on a particular college campus, plus or minus .5 years. We'd like to be 99% confident about our result. From a previous study, we know that the standard deviation for the population is 2.9 years.

Step 1: *Find $z_{\alpha/2}$ by dividing the confidence interval by two, and looking that area up in the z-table (see resources in back of book):*

$.99 / 2 = .495$

The closest z-score for .495 is **2.58**

Step 2: *Multiply Step 1 by the standard deviation.*

$2.58 \times 2.9 =$ **7.482**

Step 3: *Divide Step 2 by the margin of error:*

7.482 / .5 = ***14.96***

Step 4: *Square Step 3:*

14.96 × 14.96 = ***223.8016***

So we need a sample of ***224*** *people.*

3. *How to Find a Confidence Interval for a population*

Sample problem: 590 people applied to the Bachelor's in Elementary Education program at a certain state college. Of those applicants, 57 were men. Find the 90% confidence interval of the true proportion of men who applied to the program.

Step 1: *Read the question carefully and figure out the following variables:*

- α: subtract the given confidence interval from 1.

  1 - .9 = **.1**

- $z_{\alpha/2}$: divide the given confidence interval by 2, then look up that area in the z-table.

  .9 / 2 = .4500

  The closest z-value to an area of .4500 is **1.65**

- $\hat{p}$: divide the proportion given (i.e. the smaller number) by the sample size.

  57 / 510 = **.112**

- $\hat{q}$: Subtract $\hat{p}$ from 1.

  1 - .112 = **.888**

Step 2: *Multiply $\hat{p}$ by $\hat{q}$:*

.112 × .888 = **.099456**

Step 3: *Divide Step 2 by the sample size:*

.099456 / 510 = **.000195011765**

Step 4: *Take the square root of Step 3*:

√.000195011765 = **.0139646613**

Step 5: *Multiply Step 4 by $z_{a/2}$:*

.0139646613 × 1.65 = **.0230416911**

Step 6: *For the lower percentage, subtract Step 4 from $\hat{p}$*:

.112 - .0230416911 = .089 = **8.9%**

Step 7: *For the upper percentage, add Step 4 to $\hat{p}$*:

.112 + .0230416911 = .135 = **13.5%**

4. *How to Find a Confidence Interval for a Sample*

When you don't know anything about a population's behavior (i.e. you're just looking at data for a sample), you need to use the t-distribution table instead of the z-distribution table (see resources at the back of the book) to find the confidence interval.

Sample problem**:** A group of 10 foot surgery patients had a mean weight of 240 pounds. The sample standard deviation was 25 pounds. Find the 95% confidence interval for the true mean weight of all foot surgery patients. Assume normal distribution.

Step 1: *Subtract 1 from your sample size***.**

10 – 1 = **9**

This gives you degrees of freedom, which you'll need in Step 3.

Step 2: *Subtract the confidence level from 1, then divide by two:*

(1 – .95) / 2 = **.025**

Step 3: *Look up your answers to Step 1 and 2 in the* t-distribution table**.** For 9 degrees of freedom (df**)** and α = .025, the result is **2.262**.

df	α = 0.1	0.05	0.025	0.01	0.005	0.001	0.0005
∞	t_α=1.282	1.645	1.960	2.326	2.576	3.091	3.291
1	3.078	6.314	12.706	31.821	63.656	318.289	636.578
2	1.886	2.920	4.303	6.965	9.925	22.328	31.600
3	1.638	2.353	3.182	4.541	5.841	10.214	12.924
4	1.533	2.132	2.776	3.747	4.604	7.173	8.610
5	1.476	2.015	2.571	3.365	4.032	5.894	6.869
6	1.440	1.943	2.447	3.143	3.707	5.208	5.959
7	1.415	1.895	2.365	2.998	3.499	4.785	5.408
8	1.397	1.860	2.306	2.896	3.355	4.501	5.041
9	1.383	1.833	2.262				

Step 4: *Divide your sample standard deviation by the square root of your sample size:*

$25 / \sqrt{10} = \mathbf{7.90569415}$

Step 5: *Multiply Step 3 by Step 4:*

$2.262 \times 7.90569415 = \mathbf{17.8826802}$

Step 6: *For the lower end of the range, subtract Step 5 from the sample mean:*

$240 - 17.8826802 = \mathbf{222.117}$

Step 7: *For the upper end of the range, add Step 5 to the sample mean:*

$240 + 17.8826802 = \mathbf{257.883}$

5. How to Construct a Confidence Interval from Data Using the t-Distribution

Sample problem: Construct a 98% Confidence Interval based on the following data: 45, 55, 67, 45, 68, 79, 98, 87, 84, 82.

Step 1: *Find the mean and standard deviation for the data:*

Mean: **71**

Standard Deviation: **18.172**

For information on finding the mean, look in Chapter 1, "How to Find the Mean, Mode, and Median." For information on finding the standard deviation, look in Chapter 2, "How to Find the Sample Variance and Standard Deviation."

Step 2: *Subtract 1 from your sample size to find the degrees of freedom (df)***.** We have 10 numbers listed, so our sample size is 10:

10 – 1 = **9 degrees of freedom**

Step 3: *Subtract the confidence level from 1, then divide by two*:

(1 - .98) / 2

= .02 / 2

= **.01**

Step 4: *Look up df (Step 2) and α (Step 3) in the t-distribution table (see back of book for resources).* For df = 9 and α = .01, the table gives us **2.821**.

Step 4: *Divide your standard deviation (Step 1) by the square root of your sample size:*

18.172 / √10 = **5.75**

Step 5: *Multiply Step 3 by Step 4:*

2.2821 × 5.75 = **13.122075**

Step 6: *For the lower end of the range, subtract Step 5 from the mean (Step 1):*

71 – 13.122075 = **57.877925**

Step 7: *For the upper end of the range, add Step 5 to the mean (Step 1):*

71 + 13.122075 = **84.122**

Our 98% confidence interval is **57.877925** to **84.122**.

6. Finding Confidence Intervals for Two Populations (Proportions)

Sample problem: A study revealed that 65% of men surveyed supported the war in Afghanistan and 33% of women supported the war. If 100 men and 75 women were surveyed, find the 90% confidence interval for the data's true difference in proportions.

Step 1: *Find the following variables from the information given in the question:*

- n_1 (population sample 1): **100**
- $\hat{p}_1$ (population sample 1, positive response): 65% or **.65**
- $\hat{q}_1$ (population sample 1, negative response): 35% or **.35**
- n_2 (population sample 2): **75**
- $\hat{p}_2$ (population sample 2, positive response): 33% or **.33**
- $\hat{q}_2$ (population sample 2, negative response): 67% or **.67**

Step 2: *Find $z_{\alpha/2}$.* Divide the confidence interval by two, and look that area up in the z-table (see resources at back of book):

$.9 / 2 = .45$

$z_{\alpha/2} =$ **1.65**

Step 3: *Enter your data into the following formula and solve:*

$$(\hat{p}_1 - \hat{p}_2) - z_{a2}\sqrt{\frac{\hat{p}_1\hat{q}_1}{n_1} + \frac{\hat{p}_2\hat{q}_2}{n_2}} < \hat{p}_1 - \hat{p}_2 < (\hat{p}_1 - \hat{p}_2) - z_{\alpha 2}\sqrt{\frac{\hat{p}_1\hat{q}_1}{n_1} + \frac{\hat{p}_2\hat{q}_2}{n_2}}$$

Don't worry, here's how to solve that formula step-by-step:

1. Replace the variables with your numbers:

$$(.65-.33)-1.65\times\sqrt{\frac{.65\times.35}{100}+\frac{.33\times.67}{75}}<.65-.33<(.65-.33)-1.65\times\sqrt{\frac{.65\times.35}{100}+\frac{.33\times.67}{75}}$$

2. Multiply $\hat{p}_1$ and $\hat{q}_1$ together
   .65 × .35 = **.2275**

3. Divide your answer to (2) by n_1. Set this number aside:
   .2275 / 100 = **.00275**

4. Multiply $\hat{p}_2$ and $\hat{q}_2$ together:
   .33 × .67 = **.2211**

5. Divide your answer to (4) by n_2:
   .2211 / 75 = **.002948**

6. Add (3) and (5) together:
   .00275 + .002948 = **.005698**

7. Take the square root of (6):
   √.005698 = **.075485**

8. Multiply (7) by $z_{\alpha/2}$ found in Step 2:
   .075485 × 1.65 = **.12455**

9. Subtract $\hat{p}_2$ from $\hat{p}_1$:
   .65 - .33 = **.32**

10. Subtract (8) from (9) to get the left limit:
    .32 - .12455 = **.19545**

11. Add (8) to (9) to get the right limit:
    .32 + .12455 = **.44455**

Our 90% confidence interval is **.19545** to **.44455**. Easy!

Hypothesis Testing

1. How to find a Critical Value in 10 seconds (two-tailed test)

Let's say you have an alpha level of .05 for this example.

Step 1: *Subtract alpha from 1.*

1 - .05 = **.95**

Step 2: *Divide Step 1 by 2* (because we are looking for a two-tailed test):

.95 / 2 = **.475**

Step 3: *Look at your z-table and locate the alpha level in the middle section of the z-table*. The fastest way to do this is to use an online z-table and use the find function of your browser (usually CTRL+F). In this example we're going to look for .475, so go ahead and press CTRL+F, then type in .475.

Step 4: In this example, you should have found the number .4750. Look to the far left or the row, you'll see the number 1.9 and look to the top of the column, you'll see .06. Add them together to get **1.96**. That's the critical value!

Tip: The critical value appears twice in the z table because you're looking for both a left hand and a right hand tail, so don't forget to add plus or minus! In our example you'd get **±1.96**.

2. How to Decide whether a Hypothesis Test is a One Tailed Test or a Two Tailed Test

In a hypothesis test, you are asked to decide if a claim is true or not. For example, if someone says "all Floridians have a 50% increased chance of melanoma," it's up to you to decide if this claim holds merit. The first step is to look up a z-value, and in order to do *that*, you need to know if it's a one-tailed test or a two-tailed test. Luckily, you can figure this out in just a couple of steps.

Sample problem 1: "A government official claims that the dropout rate for local schools **is 25%.** Last year, 190 out of 603 students dropped out. Is there enough evidence to reject the government official's claim?"

Sample problem 2: "A government official claims that the dropout rate for local schools **is less than 25%.** Last year, 190 out of 603 students dropped out. Is there enough evidence to reject the government official's claim?"

Sample problem 3: "A government official claims that the dropout rate for local schools **is greater than 25%.** Last year, 190 out of 603 students dropped out. Is there enough evidence to reject the government official's claim?"

Step 2: *Rephrase the claim in the question with an equation:*

- Sample problem 1: Drop out rate = 25%
- Sample problem 2: Drop out rate < 25%
- Sample problem 3: Drop out rate > 25%

Step 3: *If Step 2 has an equal sign in it, this is a two-tailed test. If it has > or <, it's a one-tailed test.*

3. How to Decide whether a Hypothesis Test is a Left-Tailed Test or a Right-Tailed Test

In a hypothesis test that you know is one-tailed, you need to decide if it's a left-tailed test or a right-tailed test in order to figure out the z-score.

Sample hypothesis 1: Drop out rate < 25%
Sample hypothesis 2: Drop out rate > 75%

Step 1: *Draw a normal distribution curve.* Assume the area under the curve equals 100%, then shade in the related area. Choose the picture that looks most like the picture you've drawn, and then read the information underneath that picture.

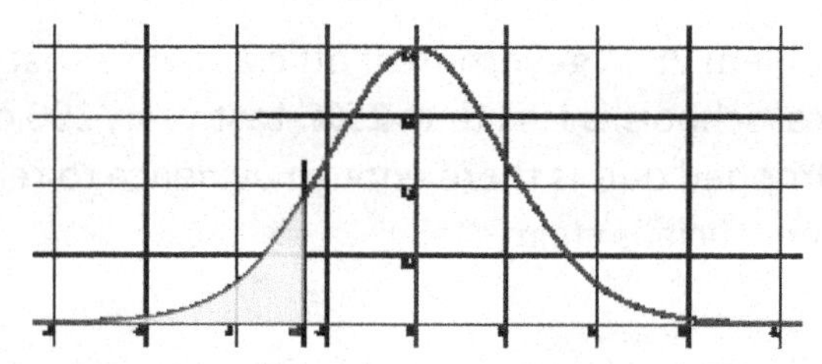

The yellow area could illustrate the area less than 25%. From this diagram you can clearly see that it is a left-tailed test (because the shaded area is on the ***left***).

The yellow area could illustrate the area greater than 75%. From this diagram you can clearly see that it is a right-tailed test (because the shaded area is on the ***right***).

4. *How to State the Null Hypothesis: Part One*

You'll be asked to convert a word problem into a hypothesis statement that will include a null hypothesis and an alternate hypothesis.

Sample problem: A researcher thinks that if knee surgery patients go to physical therapy twice a week (instead of 3 times), their recovery period will be longer. Average recovery time for knee surgery patients is 8.2 weeks.

Step 1: *Figure out the hypothesis from the problem*. The hypothesis is usually hidden in a word problem, and is sometimes a statement of what you expect to happen in the experiment. The hypothesis in the above question is "I expect the average recovery period to be greater than 8.2 weeks."

Step 2: *Convert the hypothesis to math*. Remember that the average is sometimes written as μ:

H_1: μ > 8.2

That's H_1 (the hypothesis): μ (the average) > (is greater than) 8.2

Step 3: *State what will happen if the hypothesis doesn't come true.* If the recovery time isn't greater than 8.2 weeks, there are only two possibilities, that the recovery time is equal to 8.2 weeks or less than 8.2 weeks.

H_0: μ ≤ 8.2

That's H_0 (the null hypothesis): μ (the average) ≤ (is less than or equal to) 8.2

But what if the researcher doesn't have any idea what will happen? See part two of Stating the Null and Alternate Hypothesis.

5. How to State the Null Hypothesis: Part Two

In the previous section on How to State the Null Hypothesis, I explained how to convert a word problem into a hypothesis statement if you have an idea about what the result will be (i.e. you think you know if the mean will be greater or smaller). But how do you state the null hypothesis if you have no idea about the experiment outcome?

Sample problem: A researcher is studying the effects of radical exercise program on knee surgery patients. There is a good chance the therapy will improve recovery time, but there's also the possibility it will make it worse. Average recovery time for knee surgery patients is 8.2 weeks.

Step 1: *State what will happen if the experiment doesn't make any difference.* That's the null hypothesis: nothing will happen. In this experiment, if nothing happens, then the recovery time will stay at 8.2 weeks.

$H_0: \mu = 8.2$

That's H_0 (the null hypothesis): μ (the average) = (is equal to) 8.2

Step 2: *Figure out the alternate hypothesis*. The alternate hypothesis is the opposite of the null hypothesis. In other words, what happens if our experiment makes a difference?

$H_1: \mu \neq 8.2$

That's H_1 (The alternate hypothesis): μ (the average) $\neq$ (is not equal to) 8.2

6. How to Support or Reject a Null Hypothesis

Your task is to decide whether the evidence supports a claim. If you have a P-value, or are asked to find a P-value, do not follow these instructions; instead go to the section "Null Hypothesis–P-Value."

Step 1: *State the null hypothesis and the alternate hypothesis ("the claim")*. If you aren't sure how to do this, review the section "How to State the Null and Alternate Hypothesis."

Step 2: *Find the critical value.* If you don't know how, review the section "How to find a critical value in 10 seconds."

Step 3: *Draw a normal distribution including your critical value.* Shade in the appropriate area (i.e. for a right- or left-tailed test, shade in the tail). This step is optional, it may help you visualize the problem.

Step 4: *If you already have the population standard deviation, go straight to Step 6. If you do not have the population standard deviation (i.e. you are presented with a list of data) go to Step 5*.

Step 5: *Find the sample mean ($\bar{X}$) and sample standard deviation (s) for the data.* For information on finding the mean, look in Chapter 1, "How to Find the Mean, Mode, and Median." For information on finding the standard deviation, look in Chapter 2, "How to Find the Sample Variance and Standard Deviation."

Step 6: *Use the following formula to find the z-value.* If you are given the population standard deviation σ, use it instead of the sample standard deviation, ***s***.

$$Z = \frac{\bar{X} - \mu}{s/\sqrt{n}}$$

1. Subtract the null hypothesis mean (μ–you figured this out in Step 1) from the mean of the data ($\bar{X}$ from Step 5). Set this number aside for a moment.
2. Divide the standard deviation (σ or s) by the square root of your sample size (the question either stated your sample size or you were given a certain amount n of data). For example, if thirty six children are in your sample and your standard deviation is 2, then 3/√36 = .5.
3. Divide your result from Step 1 by your result from Step 2.

Step 7: *Compare your answer from Step 6 with the α value given in the question.* If Step 6 is less than α, reject the null hypothesis, otherwise do not reject it.

7. How to Support or Reject a Null Hypothesis (Using a P-Value)

If you do not have a P-Value, instead look in the above section "How to Reject or Support a Null Hypothesis."

Step 1: *State the null hypothesis and the alternate hypothesis ("the claim")***.** If you aren't sure how to do this, see the section "How to State the Null Hypothesis" parts 1 and 2.

Step 2: *Find the critical value. If you don't know how, see the section "How to Find a Critical Value in 10 Seconds."*

Step 3: *Draw a normal distribution with your critical value.* Shade in the appropriate area (i.e. for a right- or left-tailed test, shade in the tail).

Step 4: *If you already have the population standard deviation, go straight to Step 6. If you do not have the population standard deviation (i.e. you are presented with a list of data) go to Step 5***.**

Step 5: *Find the mean ($\bar{X}$) and standard deviation (s) for the data.* For information on finding the mean, look in Chapter 1, "How to Find the Mean, Mode, and Median." For information on finding the standard deviation, look in Chapter 2, "How to Find the Sample Variance and Standard Deviation."

Step 6: *Use the following formula to find the z-value.* If you have the population standard deviation σ, use it instead of the sample population, *s.*

$$Z = \frac{\bar{X} - \mu}{s/\sqrt{n}}$$

1. Subtract the null hypothesis mean (μ–you figured this out in Step 1) from the mean of the sample data ($\bar{X}$) from Step 5).
2. Divide the standard deviation (σ or s) by the square root of your sample size (the question either stated your sample size or you were given a certain amount n of data). For example, if thirty six children are in your sample and your standard deviation is 2, then 3/√36 = .5.
3. Divide your result from Step 1 by your result from Step 2.

Step 7: *Find the P-Value by looking up your answer from Step 6 in the z-table (see resources at the back of this book).* Note: for a two-tailed test, you'll need to double this amount to get the P-Value.

Step 8: *Compare your answer from Step 7 with the α value given in the question.* If Step 7 is less than α, reject the null hypothesis, otherwise do not reject it.

8. How to Support or Reject a Null Hypothesis (for a Proportion)

If you have specific numbers instead of a proportion, see the section "How to Support or Reject A Null Hypothesis," parts 1 and 2.

Sample problem: A researcher claims that 16% of vegetarians are actually vegans. In a recent survey, 19 out of 100 vegetarians stated they were vegan. Is there enough evidence at $\alpha = .05$ to support this claim?

Step 1: *State the null hypothesis and the alternate hypothesis ("the claim").*

H_0: p = .16

H_1: p ≠ .16

If you aren't sure how to do this, follow the instructions in the section "How To State the Null Hypothesis."

Step 2: *Find the critical value. If you don't know how, see section "How to Find a Critical Value in 10 seconds."*

Step 3: *Compute* $\hat{p}$ by dividing the number of positive respondents from the number in the random sample:

19 / 100 = **.19**

Step 4: *Find p by converting the stated claim to a decimal:*

16% = **.16**

Step 5: Find 'q' by subtracting p from 1:

1 - .16 = **.84**

Step 6: *Use the following formula to calculate your test value:*

$$z = \frac{\hat{p} - p}{\sqrt{(p\,q)/n}}$$

1. Subtract p from $\hat{p}$:
   .19 - .16 = **.03**
2. Multiply p and q together, then divide by the number in the random sample:
   (.16 × .84) / 200 = **.000672**
3. Take the square root of your answer from (2):
   √.000672 = **.0259**
4. Divide your answer to (1) by your answer in (3):
   .03 / .0259 = **1.158**

Step 6: *Compare your answer from Step 5 with the α value given in the question*. If Step 5 is less than α, reject the null hypothesis, otherwise do not reject it. In this case, 1.158 is not less than our α, so we **do not reject the null hypothesis**.

9. How to Support or Reject a Null Hypothesis (for a Proportion: P-Value Method)

Sample problem: A researcher claims that more than 23% of community members go to church regularly. In a recent survey, 126 out of 420 people stated they went to church regularly. Is there enough evidence at $\alpha = .05$ to support this claim? Use the P-Value method.

Step 1: *State the null hypothesis and the alternate hypothesis ("the claim"):*

H_0: $p \leq .23$

H_1: $p > .23$ (claim)

If you aren't sure how to do this, see the section "How To State the Null Hypothesis."

Step 2: *Compute* $\hat{p}$ *by dividing the number of positive respondents from the number in the random sample*:

63 / 210 = **.3**

Step 3: *Find p by converting the stated claim to a decimal:*

23% = **.23**

Step 4: *Find q by subtracting p from 1:*

1 - .23 = **.77**

Step 5: *Use the following formula to calculate your test value:*

$$z = \frac{\hat{p} - p}{\sqrt{(p\,q)/n}}$$

1. Subtract p from $\hat{p}$
   .3 - .23 = **.07**
2. Multiply p and q together, then divide by the number in the random sample:
   (.23 × .77) / 420 = **.00042**
3. Take the square root of your answer to (2):
   √.1771 = **.0205**
4. Divide your answer to (1) by your answer in (3):
   .07 / .0205 = **3.41**

Step 6: *Find the P-Value by looking up your answer from Step 5 in the z-table (see resources in the back of this book).* The z-value for 3.41 is .4997. Subtract from .500:

.5 - .4977 = **.023**

Step 7: *Compare your P-value to α.* If the P-value is less, reject the null hypothesis. If the P-value is more, keep the null hypothesis. .023 < .05, so **we have enough evidence to reject the null hypothesis and accept the claim.**

Hypothesis Testing: Two Populations

1. How to Conduct a Statistical F-Test to Compare Two Variances

Step 1: *If you are given standard deviations, go to Step 2. If you are given variances to compare, go to Step 3.*

Step 2: *Square both standard deviations to get the variances.* For example, if $\sigma_1 = 9.6$ and $\sigma_2 = 10.9$, then the variances (s_1 and s_2) would be 9.6^2 = **92.16** and 10.9^2 = **118.81**.

Step 3: *Take the largest variance, and divide it by the smallest variance.* For example, if your two variances were s_1=2.5 and s2=9.4, divide

9.4 / 2.5 = **3.76**

Linear Regression

1. *How to Construct a Scatter Plot*

A scatter plot gives you a visual clue of what is happening with your data. Scatter plots in statistics create the foundation for linear regression, where we take scatter plots and try to create a usable model using functions. There are just three steps to creating a scatter plot.

Sample problem: Create a scatter plot for the following data:

x	y
3	25
4.1	25
5	30
6	29
6.1	42
6.3	46

Step 1: *Draw a graph. Label the x- and y- axis. Choose a range that includes the maximums and minimums from the given data.* In our example, our x-values go from 3 to 6.3, so a range from 3 to 7

would be appropriate.

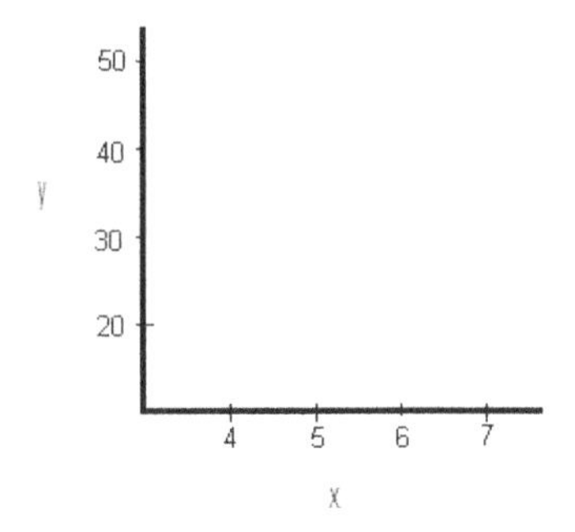

Step 2: *Draw the first point on the graph.* Our first point is (3, 25):

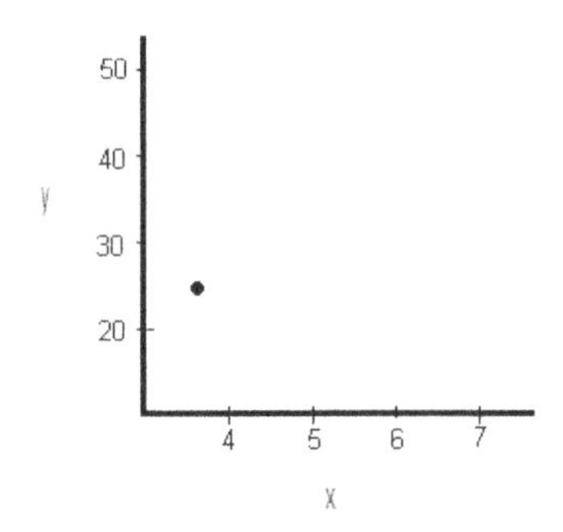

Step 3: *Draw the remaining points on the graph:*

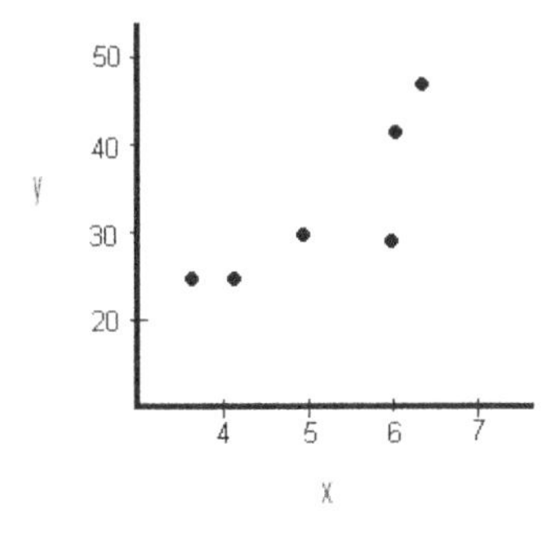

2. *How to Compute Pearson's Correlation Coefficients*

Correlation coefficients are used in statistics to measure how strong a relationship is between two variables. There are several types of correlation coefficient: Pearson's correlation is a correlation coefficient commonly used in linear regression.

Sample problem: Compute the value of the correlation coefficient from the following table:

Subject	Age x	Glucose Level y
1	43	99
2	21	65
3	25	79
4	42	75
5	57	87
6	59	81

Step 1: *Make a chart.* Use the given data, and add three more columns: xy, x^2, and y^2:

Subject	Age x	Glucose Level y	xy	x^2	y^2
1	43	99			
2	21	65			
3	25	79			
4	42	75			
5	57	87			
6	59	81			

Step 2.*Multiply x and y to fill the xy column. For example, row 1 would be 43 x 99 = 4257:*

Subject	Age	Glucose	xy	x^2	y^2
1	43	99	4257		
2	21	65	1365		
3	25	79	1975		
4	42	75	3150		
5	57	87	4959		
6	59	81	4779		

***Step 3:** Take the square of the numbers in the x column, and put the result in the x^2 column:*

Subject	Age x	Glucose Level y	xy	x^2	y^2
1	43	99	4257	1849	
2	21	65	1365	441	
3	25	79	1975	625	
4	42	75	3150	1764	
5	57	87	4959	3249	
6	59	81	4779	3481	

Step 4: *Take the square of the numbers in the y column, and put the result in the y^2 column:*

Subject	Age x	Glucose Level y	xy	x^2	y^2
1	43	99	4257	1849	9801
2	21	65	1365	441	4225
3	25	79	1975	625	6241
4	42	75	3150	1764	5625
5	57	87	4959	3249	7569
6	59	81	4779	3481	6561

Step 5: *Add up all of the numbers in the columns and put the result at the bottom.^{2} column.* The Greek letter sigma (Σ)is a short way of saying "sum of" (in other words—add them up!):

Subject	Age x	Glucose Level y	xy	x^2	y^2
1	43	99	4257	1849	9801
2	21	65	1365	441	4225
3	25	79	1975	625	6241
4	42	75	3150	1764	5625
5	57	87	4959	3249	7569
6	59	81	4779	3481	6561
Σ	**247**	**486**	**20485**	**11409**	**40022**

Step 6: *Use the following formula to work out the correlation coefficient:*

$$r = \frac{n(\sum xy) - (\sum x)(\sum y)}{\sqrt{[n(\sum x^2) - (\sum x)^2]\,[n(\sum y^2) - (\sum y)^2]}}$$

From our table:

- $\Sigma x = 247$
- $\Sigma y = 486$
- $\Sigma xy = 20{,}485$
- $\Sigma x^2 = 11{,}409$
- $\Sigma y^2 = 40{,}022$
- n is the sample size = 6

1. Replace the variables with the above values:

   $$r = \frac{6(20{,}485) - (247)(486)}{\sqrt{[6(11{,}409) - (247)^2]\,[6(40{,}022) - (486)^2]}}$$

2. Multiply n by Σxy:

   6 × 20,485 = **122,910**

3. Multiply Σx by Σy:

   247 × 486 = **120,042**

4. Subtract step (3) from step (2):

   122,910 - 120,042 = **2,868**

5. Multiply n by $\Sigma x^{2:}$

   6 × 11,409 = 68,454

6. Square Σx:

   247 × 247 = **61,009**

7. Subtract step (6) from step (5):

   68,454 – 61,009 =7,445

8. Multiply n by Σy^2:

   6 × 40,022 = **240,132**

9. Square Σy:

   486 × 486 = **236,196**

10. Subtract step (9) from (8):

    240,132 – 236,196 = **3,936**

11. Multiply step (7) by step (10):

$$7{,}445 \times 3{,}936 = 29{,}303{,}520$$

11(a). Find the square root of step (11

$$\sqrt{29{,}303{,}520} = 5413.27258$$

12. Divide step (4) by step (11):

$$2{,}868 / 5{,}413.27258 = 0.5298$$

The range of the correlation coefficient is from -1 to 1, where 1 means that the variables are completely related, -1 means the variables are completely opposite, and 0 means the variables aren't related at all. Our result is **0.5298** , or about 52.9%, which means the X and Y variables have a moderate positive correlation.

3. How to Compute a Linear Regression Test Value

Linear regression test values are used in linear regression exactly the same way as they are in hypothesis testing, but instead of working with z-tables you'll be working with t-tables.

Sample problem: Given a set of data with sample size 8 and r = .454, compute the test value.

Step 1: *Compute r, the correlation coefficient, unless it has already been given to you in the question. In our example, r =* ***.454***

Step 2: *Use the following formula to compute the test value (n is the sample size):*

$$T = r\sqrt{\frac{n-2}{1-r^2}}$$

1. *Replace the variables with your numbers:*

$$T = .454\sqrt{\frac{8-2}{1-.454^2}}$$

2. Subtract 2 from n:
   8 – 2 = **6**
3. Square r:
   .454 × .454 = **.206116**
4. Subtract step (3) from 1:
   1 - .206116 = **.793884**
5. Divide step (2) by step (4):
   6 / .793884 = **7.557779**
6. Take the square root of step (5):

   √7.557779 = **2.74914154**

7. Multiply r by step (6):

   .454 × 2.74914154 = **1.24811026**

4. *How to Find the Coefficient of Determination*

The coefficient of determination is a measure of how well a statistical model is likely to predict future outcomes. The coefficient of determination, r^2, is the square of the sample correlation coefficient between outcomes and predicted values.

The correlation coefficient for a sample is .543. What is the coefficient of determination?"

Step 1: *Find the correlation coefficient, r (it may be given to you in the question).* In our example:

r = **.543**

Step 2: *Square the correlation coefficient:*

$.543^2 = \mathbf{.295}$

Step 3: *Convert the correlation coefficient to a percentage*:

.295 = **29.5%**

5. *How to Find a Linear Regression Equation*

When a correlation coefficient shows that data is likely to be able to predict future outcomes, statisticians use linear regression to find a predictive function. Recall from elementary algebra, the equation for a line is **y = mx + b**. This article shows you how to take data, calculate linear regression, and find the equation **y' = a + bx**.

Step 1: *Make a chart of your data, filling in the columns in the same way as you would fill in the chart if you were finding the Pearson's Correlation Coefficient.* If you don't know how, see "How to Compute Pearson's Correlation Coefficient."

Subject	Age x	Glucose Level y	xy	x^2	y^2
1	43	99	4257	1849	9801
2	21	65	1365	441	4225
3	25	79	1975	625	6241
4	42	75	3150	1764	5625
5	57	87	4959	3249	7569
6	59	81	4779	3481	6561
Σ	**247**	**486**	**20485**	**11409**	**40022**

From the above table:

- $\Sigma x = 247$
- $\Sigma y = 486$
- $\Sigma xy = 20485$
- $\Sigma x^2 = 11409$
- $\Sigma y^2 = 40022$
- n is the sample size = 6

Step 2: *Calculate a using the following formula:*

$$a = \frac{(\sum y)(\sum x^2) - (\sum x)(\sum xy)}{n(\sum x^2) - (\sum x)^2}$$

1. Replace the variables with your numbers:

$$a = \frac{(486)(11409) - (247)(20485)}{6(11409) - 247^2}$$

2. Multiply Σy by Σx^2:
   486 × 11,409 = **5,544,774**
3. Multiply Σx by Σxy:
   247 × 20,485 = **5,059,795**
4. Subtract step (3) from step (2):
   5,554,774 – 5,059,795 = **484,797**
5. Multiply n by Σx^2:
   6 × 11,409 = **68,454**
6. Square Σx:
   247 × 247 = **61,009**
7. Subtract step (6) from step (5):
   68,454 – 61,009 = **7,445**
8. Divide step (4) from step (7):
   484,797 / 7,445 = **65.1171**

Step 3: *Calculate b using the following formula:*

$$b = \frac{n(\sum xy) - (\sum x)(\sum y)}{n(\sum x^2) - (\sum x)^2}$$

1. Replace the variable with your numbers:

$$b = \frac{6(20485) - (247)(486)}{6(11409) - 247^2}$$

2. Multiply n by Σxy:
   6 × 20,485 = **122,910**
3. Multiply Σx by Σy:
   247 × 486 = **120,042**
4. Subtract step (3) from step (2):
   122,910 – 120,042 = **2,868**
5. Multiply n by Σx^2:

$6 \times 11{,}409 =$ **68,454**

6. Square Σx:

   $247 \times 247 =$ **61,009**

7. Subtract step (6) from step (5):

   $68{,}454 – 61{,}009 =$ **7,445**

8. Divide step (4) by step (7):

   $2{,}868 / 7{,}445 =$ **.385225**

***Step 3**: Insert the values from steps 2 and 3 into the equation:*

$$y' = a + bx$$

$$y' = 65.1171 + .385225x$$

* **Note:** that this example has a low correlation coefficient, and therefore wouldn't be good at predicting anything.

6. How to Find a Linear Regression Slope

Step 1: *Find the following data from the information given*: Σx, Σy, Σxy, Σx^2, Σy^2. If you don't remember how to get those variables from data, see the section "How to Compute Pearson's Correlation Coefficients." Follow the Steps there to create a table and find Σx, Σy, Σxy, Σx^2, and Σy^2.

Step 2: *Insert the data into the b formula*:

$$b = \frac{n(\sum xy) - (\sum x)(\sum y)}{n(\sum x^2) - (\sum x)^2}$$

You can find more comprehensive instructions on how to work the formula in the section above, "How to Find a Linear Regression Equation."

Chi Square

1. How to Find A Critical Chi Square Value For A right Tailed Test

Chi squared distributions and the associated tables are used in significance testing. This article will tell you in a few short steps how to find a critical chi square value for a right tailed test, given degrees of freedom and the significance level.

Step 1: *Open the chi squared table* (see the resources section at the back of the book). There may be two tables listed: one table is for upper limits and one is for lower limits. You will need to use the upper limit table for the right tailed tests.

Step 2: *Find the degrees of freedom in the left hand column*. The degrees of freedom symbol is the Greek letter v.

Step 3: *Look along the top of the table to find the given significance (i.e. the probability of exceeding the critical value).* The symbol is the Greek letter α.

Step 4: *Look at the intersection of the row in Step 1 and the column in Step 2.* This is the critical chi square value.

Tip: Sometimes you will be given n (the sample size) instead of degrees of freedom. To find degrees of freedom, simply subtract 1 from this number

If n = 6

Then df = 6 – 1 = **5**

2. How to Find a Critical Chi Square Value for a Left Tailed Test

The chi squared statistic is used to compare two sets of data. To find the critical chi square value for a left tailed test, you will be using the table labeled "lower limits."

Step 1: *Subtract your significance level (α) from 1.* For example, if your significance level is .025, then:

1 - .025 = **.975**

Find this value at the top of the chi square table, heading a column.

Step 2: *Find the given degrees of freedom (v) in the left hand column.*

Step 3: *Trace down the significance level column with your finger until you find the row labeled with the given degrees of freedom.* The value you find at this intersection is the critical chi square value for the left-tailed test.

Resources

Tables

Binomial Distribution Table

http://www.statisticshowto.com/tables/binomial-distribution-table/

Chi Square Table

http://www.statisticshowto.com/tables/chi-squared-distribution-table-right-tail/

F-Table

http://www.statisticshowto.com/tables/f-table/

PPMC Critical Values

http://www.statisticshowto.com/tables/ppmc-critical-values/

T-Distribution Table

http://www.statisticshowto.com/tables/t-distribution-table/

Z-Table

http://www.statisticshowto.com/tables/z-table/

Online Calculators

Binomial Distribution Calculator

http://www.statisticshowto.com/calculators/binomial-distribution-calculator/

Interquartile Range Calculator:
http://www.statisticshowto.com/calculators/interquartile-range-calculator/

Permutation and Combination Calculator

http://www.statisticshowto.com/calculators/permutation-calculator-and-combination-calculator/

Variance and Standard Deviation Calculator

http://www.statisticshowto.com/calculators/variance-and-standard-deviation-calculator/

Glossary

!	Factorial
f	Frequency
n	Number, or sample size
Q1	First Quartile
Q3	Third Quartile
σ^2	Variance
P(X)	Probability of an event, X
E	Margin of error
E(X)	Expected value
p	Probability of success (binomial distribution)
q	Probability of failure (binomial distribution)
μ	Mean, average
σ	Standard Deviation
$\overline{X}$	Sample Mean
α	Alpha level, significance level
H_1	Alternate hypothesis
s	Sample variance estimate
r	Correlation coefficient
r^2	Coefficient of determination
Σ	Sum of
a	Y intercept (linear regression)
b	Slope of the line (linear regression)
v	Degrees of freedom

The Practically Cheating Statistics Handbook

TI-83 Companion Guide

S. Deviant, MAT

StatisticsHowTo.com

info@statisticshowto.com

Contents

Basics

Mean and Median

Finding the mean or median from a list of data can be accomplished in two ways: by entering a list of data, or by using the home screen to type the commands. Using the list feature is just as easy as entering the data onto the home screen, and it has the added advantage that you can use the data for other purposes after you have calculated your mean and median (for example, you might want to create a histogram).

Sample problem: Find the mean and the median for the height of the top 20 buildings in NYC. The heights, (in feet) are: 1250, 1200, 1046, 1046, 952, 927, 915, 861, 850, 814, 813, 809, 808, 806, 792, 778, 757, 755, 752, and 750.

Step 1: Enter the above data into a list. Press the STAT button and then press ENTER. Enter the first number (1250), and then press ENTER. Continue entering numbers, pressing the ENTER button after each entry.

Step 2: Press the STAT button.

Step 3: Press → to highlight **Calc**.

Step 4: Press ENTER to choose **1-Var Stats**.

Step 5: Press ENTER again. The calculator will return the mean, x. For this list of data, the mean is **877.667** feet (rounded to 3 decimal places). The standard deviation is Sx=**147.022**.

Step 6: Arrow down until you see **Med**. This is the median; for the above data, the median is **813** feet.

Interquartile Range

The interquartile range is the distance between the top of the lower quartile and the bottom of the top quartile.

Sample problem: Find the interquartile range for the heights of the top 10 buildings in the world. The heights, (in feet) are: 2717, 2063, 2001, 1815, 1516, 1503, 1482, 1377, 1312, 1272.

Step 1: Enter the above data into a list. Press the STAT button and then press ENTER. Enter the first number (2717), and then press ENTER. Continue entering numbers, pressing ENTER after each entry.

Step 2: Press STAT.

Step 3: Press → (the arrow keys are located at the top right of the keypad) to select **Calc**.

Step 4: Press ENTER to highlight **1-Var Stats**.

Step 5: Press ENTER again to bring up a list of stats.

Step 6: Scroll down the list with the arrow keys to find Q1 and Q3. Subtract Q3 from Q1 to find the **interquartile range** (which in this set of numbers would be **624** feet).

Variance

The variance is simply the standard deviation squared. You could find the standard deviation for a list of data and square the result, but you won't get an accurate answer unless you square the entire answer, including *all* of the significant digits (not, for example, just the two I rounded to in the standard deviation question). Here's how to get the TI-83 variance without you having to perform a manual calculation:

Sample problem: Find the variance for the heights of the top 12 buildings in London, England. The heights, (in feet) are: 800, 720, 655, 655, 625, 600, 590, 529, 513, 502, 502, 502.

Step 1: Enter the above data into a list. Press the STAT button and then press ENTER. Enter the first number (800), and then press ENTER. Continue entering numbers, pressing ENTER after each entry.

Step 2: Press STAT. Press → to select **Calc**.

Step 3: Press ENTER to highlight **1-Var Stats**. Press ENTER again to bring up a list of stats.

Step 4: Press VARS 5 to bring up a list of the available Statistics variables.

Step 5: Press 3 to select "Sx" which is our standard deviation.

Step 6: Press x^2, then ENTER to display the variance, which is **9326.628788**.

Standard Deviation

Standard deviations give us an idea of how much data is contained within a certain part of a distribution curve. This information is especially useful when using normal distribution curves (which you'll, no doubt, learn about during your statistics course).

Sample problem: Find the standard deviation for the heights of the top 12 buildings in London, England. The heights, (in feet) are: 800, 720, 655, 655, 625, 600, 590, 529, 513, 502, 502, 502.

Step 1: Enter the above data into a list. Press the STAT button and then press ENTER. Enter the first number (800), and then press ENTER. Continue entering numbers, pressing ENTER after each entry.

Step 2: Press STAT.

Step 3: Press → (the arrow keys are located at the top right of the keypad) to select **Calc**.

Step 4: Press ENTER to highlight **1-Var Stats**.

Step 5: Press ENTER again to bring up a list of stats. The standard deviation (Sx) for the above list of data is **96.57** feet (rounded to 2 decimal places).

Graphs

Box Plot

Sample problem: You have a list of IQ scores for a gifted classroom in a particular elementary school. The IQ scores are: 118, 123, 124, 125, 127, 128, 129, 130, 130, 133, 136, 138, 141, 142, 149, 150, 154. That list doesn't tell you much about anything. Create a box plot to make sense of this data.

Step 1: Press STAT ENTER, to edit L1.

Step 2: Enter the data from the problem into the list:

1 1 8 ENTER
1 2 3 ENTER
1 2 4 ENTER
1 2 5 ENTER
1 2 7 ENTER
1 2 8 ENTER
1 2 9 ENTER
1 3 0 ENTER
1 3 0 ENTER
1 3 3 ENTER
1 3 6 ENTER
1 3 8 ENTER
1 4 1 ENTER
1 4 2 ENTER
1 4 9 ENTER
1 5 0 ENTER
1 5 4 ENTER

Step 3: Press 2ND Y=, to access the **Stat Plot** menu.

Step 4: Press ENTER ENTER to turn on **Plot1**.

Step 5: Arrow down to **Type**, which has 6 icons to the right of it. Highlight the bottom middle icon, which looks like a syringe with two plungers, and press ENTER to select it.

Step 6: Make sure the **XList** entry reads "L_1". If it doesn't, arrow down to it, Press CLEAR 2ND 1.

Step 7: Press GRAPH. You should see your Box plot!

Tip: If when you press GRAPH, you see the message "Err: Stat", or you just don't see a box plot like you expect to, then press WINDOW, and try different settings. Especially try changing the **Xscl** (X Scale) item to a larger value.

Cumulative Frequency Table

A cumulative frequency table shows shows the sum of data up to a specified point. A TI 83 can draw a cumulative frequency table, using values already placed in a list. Just type in the numbers, press a button, and the calculator does the rest.

Sample problem: Construct a cumulative frequency table for the grades of school children in Rock Hill elementary school (grades k-5). The grades are A (90+), B (80+), C (70+), and D (60+). The number of students achieving those grades are: A (20), B (21) C (52) and D (29).

Step 1: Enter the data list L1. Press the STAT button and then press ENTER. Press:

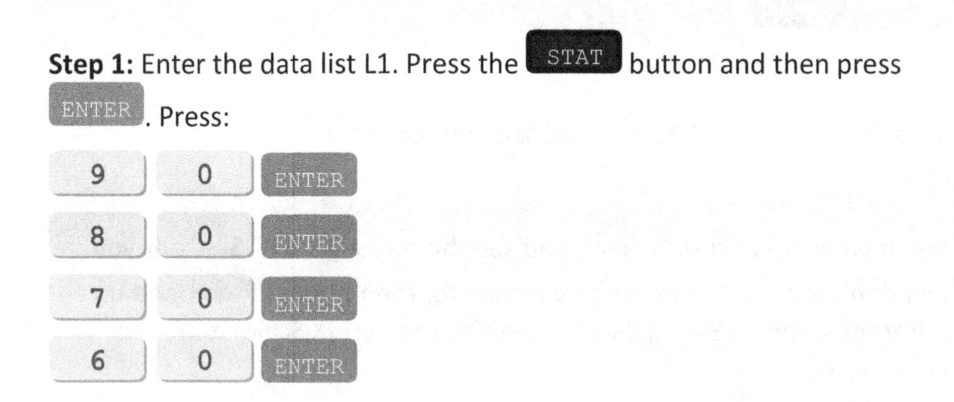

Step 2: Scroll to the right to list L2. Enter the next column's values:

Step 3: Scroll to the right and up to highlight L3.

Step 4: Press 2ND STAT to reach the **LIST** menu.

Step 5: Scroll to the right to highlight **OPS** then press 6 for **cumSum**.

Step 6: Press 2ND 1 to get L1, then press).

Step 7: Press ENTER. The TI-83 will sum the cumulative data in the L2 column and replace the results in L3, giving you a cumulative distribution table.

Tip: To clear the data from a list, place the cursor on the list name (for example, L1), press CLEAR, and then ENTER.

Frequency Chart or Histogram

A histogram is one of the most popular ways the display a frequency distribution. A series of vertical rectangles fit together on a chart. The histogram is an important tool for exploratory data analysis, and can be plotted on a TI 83 graphing calculator in less time than it would take you to get out a sheet of paper, a pencil, and a ruler.

Sample problem: draw a histogram for the following recent test scores in a statistics class: 45, 67, 68, 69, 74, 76, 75, 77, 79, 84, 86, 90.

Step 1: Press STAT ENTER, to edit L1.

Step 2: Enter the data from the problem into the list:

4 5 ENTER
6 7 ENTER
6 8 ENTER
6 9 ENTER
7 4 ENTER
7 6 ENTER
7 5 ENTER
7 7 ENTER
7 9 ENTER
8 4 ENTER
8 6 ENTER
9 0 ENTER

Step 3: Press 2ND Y=, to access the **Stat Plot** menu.

Step 4: Press ENTER ENTER to turn on **Plot1**.

Step 5: Arrow down to **Type**, which has 6 icons to the right of it. Highlight the top right icon, which looks like a histogram, and press ENTER to select it.

Step 6: Make sure the **XList** entry reads "L_1". If it doesn't, arrow down to it, CLEAR 2ND 1.

Step 7: Press GRAPH. You should see your Histogram!

Tip: If when you press GRAPH, you see the message "Err: Stat", or you just don't see a histogram like you expect to, then Press WINDOW, and try different settings. Especially try changing the **Xscl** (X Scale) item to a larger value.

Scatter Plot

In order to graph a TI-83 scatter plot, you'll need a set of "bivariate" data. Bivariate data means you have both X and Y values, for example, weight (your X) and height (your Y). The XY values can be in two separate lists, or they can be written as XY cordinates (x,y).

Sample problem: Create a TI-83 scatter plot for the following coordinates (2,3), (4,4), (6,9), (8,11), and (10,12).

Step 1: Press STAT ENTER to enter the lists screen. If you already have data in L1 or L2, clear the data: move to cursor onto L1, press CLEAR and then ENTER. Repeat for L2.

Step 2: *Enter your x-variables, one at a time.* Follow each number by pressing the ENTER key. For our list, you would enter:

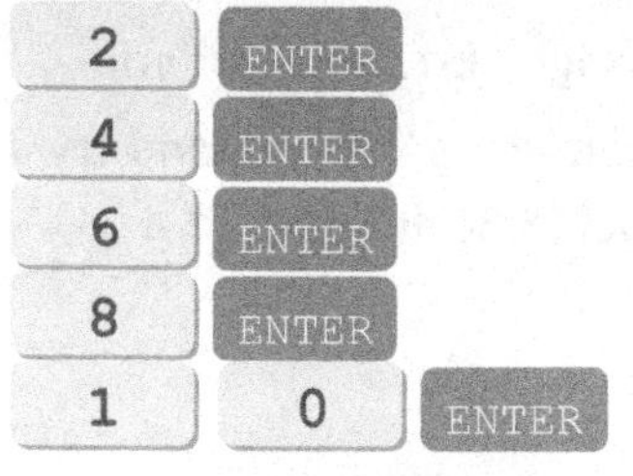

Step 3: Use the → key to scroll across to the next column, L2.

Step 4: Enter your y-variables, one at a time. Follow each number by pressing the enter key. For our list, you would enter:

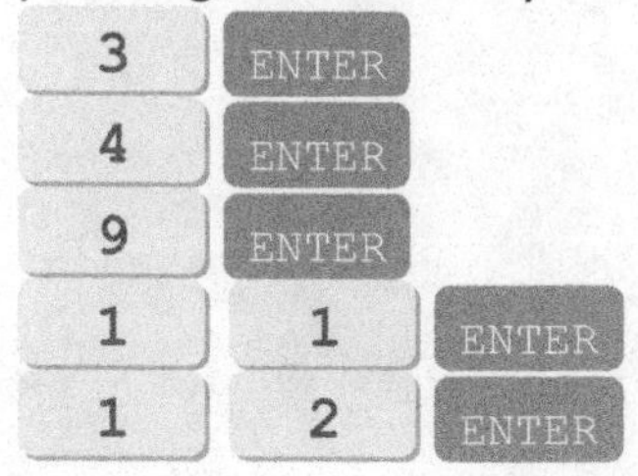

Step 5: Press 2ND and then press Y= (Statplot).

Step 6: Press ENTER to enter StatPlots for Plot1.

Step 7: Press ENTER to turn Plot1 "ON."

Step 8: Arrow down to the next line ("Type") and highlight the scatter plot (the first image).

Step 9: Arrow down to "Xlist." If "L1" isn't showing, press 2ND and 1. Arrow down to "Ylist." If "L2" isn't showing, press 2ND and 2.

Step 10: Press ZOOM and 9. This should bring up a scatter plot on your screen.

Tip: Hit TRACE and press the right and left arrow buttons to move from point to point, displaying the XY values for those points.

Binomial Probability Distribution

Combinations

Sample problem: If there are 5 people, Barb, Sue, Jan, Jim, and Rob, and only three will be chosen for the new Parent Teacher Association, how many combinations are possible for the committee?

Step 1: Enter the number of items on the home screen (the main input area) that you have to choose from. In the above example, you have 5 people, so press 5.

Step 2: Press the MATH button.

Step 3: Press → to tab across to **PRB**.

Step 4: Press 3. This inserts **nCr** onto the home screen.

Step 5: Enter the number of items you want to choose. In the above example, you want to choose 3, so press 3.

Step 6: Press ENTER. The calculator will return the result: **10**.

Permutations

Sample problem: If there are 5 people, Barb, Sue, Jan, Jim, and Rob, and three will be chosen for the new Parent Teacher Association. The first person picked will be the president, then the vice president, then the secretary. How many permutations are possible for the committee?

Step 1: Enter the number of items on the home screen (the main input area) that you have to choose from. In the above example, you have 5 people, so press 5.

Step 2: Press the MATH button.

Step 3: Press → to tab across to **PRB**.

Step 4: Press 2. This inserts **nPr** onto the home screen.

Step 5: Enter the number of items you want to choose. In the above example, you want to choose 3, so press 3.

Step 6: Press ENTER. The calculator will return the result: **60**.

Mean and Standard Deviation for a Binomial Probability Distribution

Binomial probability distributions are used when dealing with two outcomes from a fixed number of independent trials. As a simple example, you might want to figure out the probability of a voter voting yes or no on a certain ballot proposition. P(x) can quickly be calculated on the TI 83 graphing calculator, which has all of the binomial probability tables built into it.

Sample problem: Find the mean and standard deviation for a binomial distribution with n = 5 and p = 0.12.

Mean

Step 1: Multiply n by p:

5 × . 1 2 ENTER

= .6

Hey, that was easy!

Standard Deviation

Step 1: Subtract p from 1 to find q.

1 – . 1 2 ENTER

= **.88**

Step 2: Multiply n times p times q.

5 × . 1 2 × . 8 8 ENTER

= **.528**

Step 3: Find the square root of the answer from Step 2.

2nd x^2 . 5 2 8 ENTER

.727 (rounded to 3 decimal places).

Bionomial Probability: BinomCDF

The TI-83 BinomPDF and TI-83 BinomCDF functions can assist you in solving binomial probability questions in seconds. The difference between the two functions is that one (BinomPDF) is for a single number (for example, three tosses of a coin), while the other (BinomCDF) is a cumulative probability (for example, 0 to 3 tosses of a coin).

Sample problem: Bob's batting average is .321. If he steps up to the plate four times, find the probability that he gets up to three hits.

Step 1: Press 2ND VARS to get the **DISTR** option. Scroll down to **A:BinomCDF** and then press ENTER.

Step 2: Enter the number of trials. Bob bats four times, so the number of trials is **4**. Press 4 followed by ,.

Step 3: Enter the Probability of Success, **P**. Bob has a batting average of **.321**, so enter . 3 2 1, then press ,.

Step 4: Enter the X value. We want to know what the probability is that Bob will get up to three hits, so enter 3 and then press).

Step 5: Press ENTER for the result. Bob's chances of getting up to three hits is **.989**.

Tip: Instead of scrolling to **A:BinomCDF** you can press ALPHA MATH.

Bionomial Probability: BinomPDF

Sample problem: Sue has a batting average of .270. If she is at bat three times, what is the probability that she gets exactly two hits?

Step 1: Press 2ND VARS to get the **DISTR** option. Press 0 for **binomPDF**.

Step 2: Enter the number of trials from the question. Sue bats three times, so the number of trials is **3**. Press 3 and hit the , key.

Step 3: Enter the Probability of Success, **P**. Sue's batting average is **.240**, so enter . 2 7 0, then hit the , key.

Step 4: Enter the X value. We want to know Sue's chances of getting exactly two hits, so enter 2 and then press).

Step 5: Press ENTER for the result. Sue's chances of getting exactly two hits is **.159651**.

Warning: **BinomialPdf** is an exact probability for one value of x. If you want to find a cumulative probability (for example, what are John's chances of getting 0 or 1 hits?) you will need the use the BinomialCdf function.

Normal Distributions

Central Limit Theorem

The TI 83 calculator has a built in function that can help you calculator probabilities of central theorem word problems. The function, normalcdf, requires you to enter a lower bound, upper bound, mean, and standard deviation.

Sample problem: A fertilizer company manufactures organic fertilizer in 10 pound bags with a standard deviation of 1.25 pounds per bag. What is the probability that a random sample of 150 bags will have a mean between 9 and 9.5 pounds?

Step 1: Press 

Step 2: Enter your variables (lower bound, upper bound, mean, and standard deviation): 9 , 9 . 5 , 1 0 , (1 . 2 5 ÷ 2nd x^2 1 5 0)).

Step 3: Press ENTER. This returns the probability of **.159**, or **15.9%**.

Tip: Sampling distributions require that the standard deviation of the mean is σ / √(n), so make sure you enter that as the standard deviation.

Normally Distributed Probability

Sample problem: A group of students taking end of semester statistics exams at a certain college have a mean score of 75 and a standard deviation of 5 points. What is the probability that a given student will score between 90 and 100 points?

Step 1: Press 2ND VARS 2 to get **normalcdf**.

Step 2: Enter 9 0 for the lower bound, followed by , . Then enter 1 0 0 for the upper bound, followed by , .

Step 3: Press 7 5 (for the mean), followed by , and then 5 (for the standard deviation).

Step 4: Close the argument list with a) . Your display should now read **normalcdf(90, 100, 75, 5)**. Now press ENTER. The calculator returns the probability, which in this case is **0.00135**, or **1.35%** (rounded to two decimal places).

Normal Probability Density Function

The TI-83 **normalPDF** function, accessible from the **DISTR** menu will calculate the normal probability density function, given the mean (μ) and standard deviation (σ).

Sample Problem: Graph a bell curve with a mean of 100 and standard deviation of 15.

Step 1: Press .

Step 2: Press to get **normalPDF**.

Step 3: Press .

Step 4: Press .

Step 5: Change the window values to the following:
Xmin=0
Xmax=150
Xscl=1
Ymin=0
Ymax=.05

Step 6: Press GRAPH to see your graph.

Z Scores

Find a Critical Value

A critical value in hypothesis testing separates the region where the hypothesis will be rejected from the region where the hypothesis will not be rejected. With a little arithmetic, you can look up a critical value in a z-table, or you can use the **InvNorm** function on the TI-83 graphing calculator.

Sample problem: An end of semester exam is normally distributed with a mean of 85 and a standard deviation of 10. Find the z score at the 90th percentile.

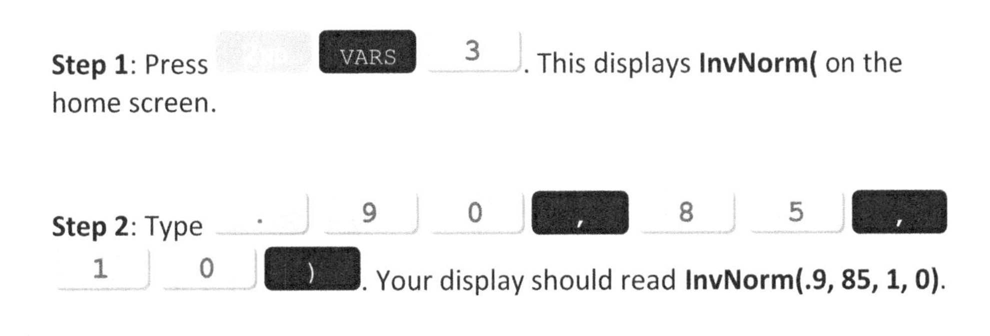

Step 1: Press . This displays **InvNorm(** on the home screen.

Step 2: Type . Your display should read **InvNorm(.9, 85, 1, 0)**.

Step 3: Press ENTER. This returns **97.82**. That means that 90% of students will have scores below 97.82.

Tip: The first entry on **InvNorm** should be a number between 0 and 1.

Hypothesis Test on a Mean

Researchers use data from a population sample to figure out when a hypothesis is true or not. If the sample data does not agree with the hypothesis, then the hypothesis is rejected. The null hypothesis (H_0) states that the observations occur by chance. The alternate hypothesis (H_1) is just the opposite, and states that the observations were influenced by a non-random cause.

Sample problem: A sample of 200 people has a mean age of 21 with a population standard deviation (σ) of 5. Test the hypothesis that the population mean is 18.9 at $\alpha = 0.05$.

Step 1: State the null hypothesis. In this case, the null hypothesis is that the population mean is 18.9, so we write:
H_0: $\mu = 18.9$

Step 2: State the alternative hypothesis. We want to know if our sample, which has a mean of 21 instead of 18.9, really is different from the population, therefore our alternate hypothesis:
H_1: $\mu \neq 18.9$

Step 3: Press STAT then press → → to select **TESTS**.

Step 4: Press 1 to select **1:Z-Test….** Press ENTER.

Step 5: Press → to select **Stats**.

Step 6: Enter the data from the problem:
μ_0: 18.9
σ: 5
x: 21
n: 200
μ: $\neq\mu_0$

Step 7: Arrow down to **Calculate** and press ENTER. The calculator shows the p-value:
$p = 2.87 \times 10^{-9}$

This is smaller than our alpha value of .05. That means we should **reject the null hypothesis**, the average age of the population is *not* 18.9 years old.

Confidence Intervals

Confidence Intervals for a Proportion

Confidence intervals are one way you can decide how accurate data is. The higher the confidence, the better the estimate (a small range of possible values): a 98% confidence interval is going to give you a better estimate of the unknown parameter than a 90% confidence interval.

Sample problem: For x = 20, and n = 24, construct a 98% confidence interval for p, the true population proportion.

Step 1: Press STAT.

Step 2: Press → to highlight **TESTS**.

Step 3: Arrow down to **A:1–PropZInt...** and then press ENTER.

Step 4: Enter your x-value: 2 0.

Step 5: Arrow down and then enter your n value: 2 4.

Step 6: Arrow down to **C-Level** and enter . 9 8.

Step 7: Arrow down to C**alculate** and press ENTER. The calculator will return the range **(.65636, 1.0103)**.

Tip: Instead of arrowing down to select **A:1–PropZInt...**, press ALPHA and MATH instead.

Confidence Interval for a Mean (Known Standard Deviation)

The only difference between having statistics such as mean, or standard deviation, versus the raw data, is that you will have to enter the data into a list in order to perform the calculation.

Sample problem: 40 items are sampled from a normally distributed population with a sample mean (x) of 22.1 and a population standard deviation (σ) of 12.8. Construct a 98% confidence interval for the true population mean.

Step 1: Press STAT, then → to **TESTS**. Press ENTER.

Step 2: Press 7 for **Z Interval**.

Step 3: Arrow over to **Stats** on the **Inpt** line and press ENTER to highlight and move to the next line, σ.

Step 4: Enter 1 2 . 8, then arrow down to x.

Step 5: Enter 2 2 . 1, then arrow down to "n."

Step 6: Enter 4 0, then arrow down to "C-Level."

Step 7: Enter . 9 8. Arrow down to **Calculate** and then press ENTER. The calculator will give you the result of (17.392, 26.808) meaning that your 98% confidence interval is **17.392 to 26.808**.

T Scores

How to Find a T Distribution

The T distribution (sometimes called Student's T distribution) is actually a set of distributions, differentiated by the degrees of freedom (df). Take a look at a traditional textbook T table, and you'll actually find *many* T tables, which can be a little overwhelming. Instead of poring over tables, you can use a TI 83 graphing calculator to assist you in finding T distribution values. You might be asked to find the area under a T curve, or (like Z scores), you might be given a certain area and asked to find the T score.

Sample problem: Find the area under a T curve with degrees of freedom 10 for P(1 ≤ X ≤ 2).

Step 1: Press to select **tcdf(**.

Step 2: Enter the lower, and upper bounds, and the degrees of freedom:

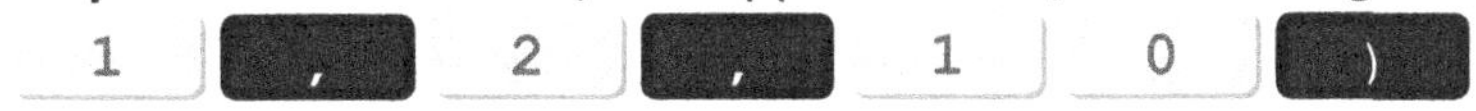

Your screen should now read **tcdf(1,2,10)**

Step 3: Press ENTER. The answer is **.133752549**, or about **13.38%**.

Linear Regression

Linear Regression

You can perform linear regression with a TI-83 in the time it takes you to input a few variables into a list. Linear regression will only give you a reasonable result if your data looks like a line on a scatter plot, so before you find the equation for a linear regression line you may want to view the data on a scatter plot first. See the Graphs chapter in this book to find out how to create a scatter plot.

Sample problem: Find a linear regression equation (of the form y = ax + b) for x-values of 1, 2, 3, 4, 5 and y-values of 3, 9, 27, 64, and 102.

Step 1: Press STAT ENTER to enter the lists screen. If you already have data in L1 or L2, clear the data: move to cursor onto L1, press CLEAR and then ENTER. Repeat for L2.

Step 2: *Enter your x-variables, one at a time.* Press:

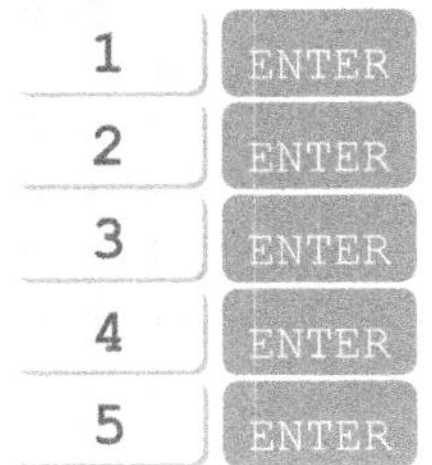

Step 3: Use the → key to scroll across to the next column, L2.

Step 4: Enter your y-variables, one at a time. Press:

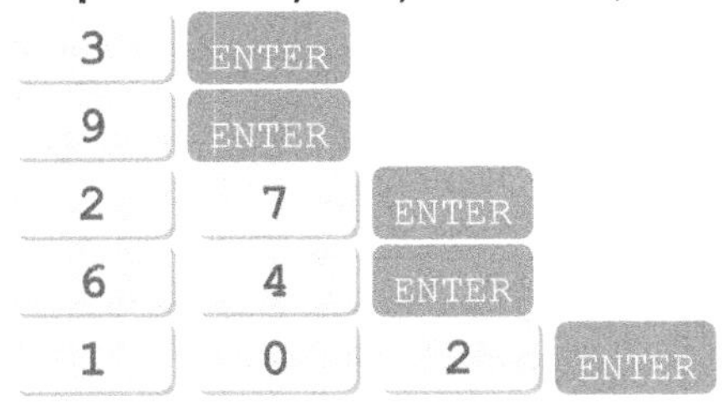

Step 5: Press the STAT button, then use the scroll key to highlight **CALC**. Press ENTER.

Step 6: Press 4 to choose **LinReg(ax+b)**. Press ENTER. The TI-83 will return the variables needed for the equation. Just insert the given variables (a, b) into the equation for linear regression (y=ax+b). For the above data, this is **y = 25.3x – 34.9**.

Two Populations

Confidence Intervals for Two Populations

Statistics about two populations is incredibly important for a variety of research areas. For example, if there's a new drug being tested for diabetes, researchers might be interested in comparing the mean blood glucose level of the new drug takers versus the mean blood glucose level of a control group. The confidence interval (CI) for the difference between the two population means is used to assist researchers in questions such as these. The TI 83 allows you to find a CI for the difference between two means in a matter of a few keystrokes.

Sample problem: Find a 98% confidence interval for the difference in means for two normally distributed populations with the following characteristics:

$x_1 = 88.5$
$\sigma_1 = 15.8$
$n_1 = 38$

$x_2 = 74.5$
$\sigma_2 = 12.3$
$n_1 = 48$

Step 1: Press STAT, then → highlight **TESTS**.

Step 2: Press 9 to select **2-SampZInt….**

Step 3: Enter the values from the problem into the appropriate rows, using the down arrow to switch between rows as you complete them.

Step 4: Press ↓ to select **Calculate** then press ENTER. The answer displayed is **(6.7467, 21.253)**. We're 98% sure that the difference between the two means is between 6.7467 and 21.253.

Made in the USA
Las Vegas, NV
13 September 2023

77533582R00105